# The Associational Tourism Economy

This timely and innovative book provides a comprehensive introduction to the concept of the associational tourism economy, exploring the collaborative nature of tourism as an interconnected system, with a focus on tourism clusters, networks, and ecosystems.

When tourists travel, they interact with a complex web of businesses, organisations, and individuals who can act independently or with a greater or lesser level of collaboration among themselves. This interdependence emphasises that the associational tourism economy is not merely about the mobility of people but also concerns interconnected socio-economic systems that shape where and why people travel. Replete with illuminating tables and diagrams, this book lays a foundation for understanding the complex, associational nature of the tourism economy from both business and policy perspectives. An extensive overview of the evolution of theoretical concepts of associational economics provides the framework for a clear breakdown of the intricate collaborative, competitive, and interpersonal relations that make up the economics of tourism, with insights into how these relations can be built and fostered to promote sustainable tourism and boost local economies.

This book will be of pivotal interest to students, scholars, and academics of tourism and hospitality research, tourism planning and development, tourism management, and tourism geography.

**Alexander Safonov** holds a PhD in Marketing and is a professional daredevil. He brings a wealth of knowledge from his diverse background in bold and unconventional adventures. His interests span the associational tourism economy and a wide range of tourism-related topics.

# Routledge Focus on Tourism and Hospitality

*Routledge Focus on Tourism and Hospitality* presents small books on big topics and how they intersect with the world of tourism and hospitality research. The idea is to fill the gap between journal article and book. This new short form series offers both established and early-career academics the flexibility to publish cutting-edge commentary on key areas of tourism and hospitality, topical issues, policy-focused research, analytical or theoretical innovations, a summary of the key players or short topics for specialized audiences in a succinct way.

**Managing People in Commercial Kitchens**
A Contemporary Approach
*Charalampos Giousmpasoglou, Evangelia Marinakou, Anastasios Zopiatis and John Cooper*

**Smart Tourism Destination Governance**
Technology and Design-Based Approach
*Tomáš Gajdošík*

**Solo Travel, Tourism and Loneliness**
A Critical Sociology
*Hugues Séraphin and Maximiliano E. Korstanje*

**The Future of Blockchain in Tourism and Hospitality**
Global Insights
*Fatima Zahra Fakir and Erdem Baydeniz*

**The Associational Tourism Economy**
An Introduction to Collaboration – Clusters, Networks, Ecosystems
*Alexander Safonov*

For more information about this series, please visit: www.routledge.com/Routledge-Focus-on-Tourism-and-Hospitality/book-series/FTH

# The Associational Tourism Economy

An Introduction to Collaboration –
Clusters, Networks, Ecosystems

**Alexander Safonov**

LONDON AND NEW YORK

First published 2025
by Routledge
4 Park Square, Milton Park, Abingdon, Oxon OX14 4RN

and by Routledge
605 Third Avenue, New York, NY 10158

*Routledge is an imprint of the Taylor & Francis Group, an informa business*

© 2025 Alexander Safonov

*British Library Cataloguing-in-Publication Data*
A catalogue record for this book is available from the British Library

ISBN: 978-1-032-27658-8 (hbk)
ISBN: 978-1-032-27659-5 (pbk)
ISBN: 978-1-003-29360-6 (ebk)

DOI: 10.4324/9781003293606

Typeset in Times New Roman
by SPi Technologies India Pvt Ltd (Straive)

To those who never give up, to those who work hard, and to the ones behind every weak man. This book is dedicated to you, the one whose unconditional love, eternal patience, and wholehearted support bring joy and meaning to my life.

# Contents

# Figures

# Tables

# Preface

Tourism is a global phenomenon deeply rooted in the history of human mobility. Since ancient times, people have travelled for various purposes such as gathering food, trade, religious pilgrimages, or exploration of new territories, establishing a tradition that laid the foundations for modern tourism. Today tourism has expanded dramatically in both scale and scope. The numbers of tourists, destinations, and means of travel have grown exponentially, driven by increased incomes, advanced transportation, and a broad array of services. Despite its evolution, however, the essence of tourism remains the same – the movement of people across space and time, driven by curiosity, necessity, or leisure.

Tourism operates as a complex system involving tourists, environments, organisations, and diverse businesses, from individual souvenir craftsmen to travel agencies, airlines, hotels, restaurants, and entertainment venues. Destinations are shaped by networks of businesses that collectively define their appeal, meaning that each business contributes to the overall destination experience. Each decision to travel initiates a chain reaction that activates numerous interdependent businesses, highlighting tourism's systemic nature. From purchasing tickets to booking accommodations and exploring local attractions, every transaction contributes to the vast, interconnected tourism economy.

The nature of tourism is inherently associational from the perspectives of both tourists and businesses. Tourists are often motivated by a desire to connect with certain social groups, experiences, or localities, whether that means visiting exclusive destinations, engaging in local culture, or sharing in the experiences of others. From a business perspective, tourism is also associational, since individual businesses rely on collaborative relationships with others to create a compelling and competitive destination. No single business can fully represent or sustain a destination alone, and even large businesses depend on the broader tourism ecosystem. Therefore, tourism businesses operate within the broader associational tourism economy, so that even when operating independently, they collectively enhance a destination's capacity to attract and retain tourists.

This inherent associational aspect highlights the unique economic and social dynamics within tourism. Tourists when travelling to destinations often seek a sense of belonging through past experiences, recommendations, shared identities, or communities associated with those places. This interdependence emphasises that the associational tourism economy is not merely about the mobility of people but also concerns interconnected socio-economic systems that shape where and why people travel.

This book introduces readers to the concept of the associational tourism economy, exploring the collaborative nature of tourism as an interconnected system with a focus on tourism clusters, networks, and ecosystems. It lays a foundation for understanding the complex, associational nature of the tourism economy.

*Alexander Safonov*

# Abbreviations

**DMO**    Destination Marketing Organisation
**ICT**    Information and Communication Technologies
**LEED**    Local Employment and Economic Development Programme
**NTO**    National Tourism Organisation
**OECD**    Organization for Economic Cooperation and Development
**RTO**    Regional Tourism Organisation
**UNWTO**    UN World Tourism Organization

# 1 The associational tourism economy

## Introduction

The consumption structure of modern societies is shifting towards services (Sorbe et al., 2018). Tourism, representing a range of services, positively impacts regions' socio-economic development since it involves many independent entities in economic activity. According to the UN World Tourism Organization (UNWTO), tourism services are an essential factor in socio-economic development, as evidenced by the growing dynamics of tourism services over the past decades. Tourism businesses significantly impact local economies and play a meaningful role in poverty reduction and employment creation (UNWTO, 2013). Moreover, tourism impacts rural economies, providing an opportunity to enrich economic activity by sharing local culture with tourists (Michael, 2007a). In almost half of the developing countries, tourism occupies a leading position in exports, influencing employment and opening new economic opportunities (UNWTO, 2013). Tourism keeps expanding and diversifying, thus staying as one of the world's largest and fastest-growing economic sectors (UNWTO, 2023).

In the economy, markets and hierarchies have long dominated the field as efficient coordinating mechanisms focusing on individual firms and their internal capacities (Powell, 1991). The emphasis has been on the idiosyncratic competitive advantage of a firm, with the assumption that the pursuit of self-interest results in efficient economic exchange. Economists have considered associations as an inefficient means of allocating scarce resources, placing greater emphasis on market relationships (Streeck & Schmitter, 1991). However, market prices fail to capture and manage the complex and ever-evolving nature of business relationships. Similarly, hierarchies, with their rigid control mechanisms, often struggle to adapt to the rapid changes (Powell, 1991).

While formal networks offer a more collaborative framework, they often represent contractual arrangements with a fee-based system, which can sometimes reduce overall effectiveness (Phillipson et al., 2006). Such

DOI: 10.4324/9781003293606-1

limitations become particularly evident in tourism, in which a detachment between where knowledge resides and where initiatives are proposed frequently exists, suggesting that successful tourism development requires closer collaboration among stakeholders. The shift towards a networking economy views businesses not in isolation but as part of a network of interdependent firms (Cooke & Morgan, 1993; De Man, 2004; Johansson et al., 2012; Molenaar, 2020). In this dynamic landscape, the state plays a crucial role in fostering conditions that enable firms, intermediary associations, and public organisations to engage in a self-organised process of interactive learning (Cooke & Morgan, 1998). Since tourism comprises micro- to medium-sized businesses that often face constraints in terms of financial resources and access to knowledge (Hall & Michael, 2006; Michael, 2007b; Novelli et al., 2006), associative thinking becomes essential for survival and growth (Kim & Shim, 2018).

The tourism economy intertwines various modes of coordination. Market dynamics are integral to tourism development, driving competition among businesses and destinations. Networks too have long been essential in tourism business activities. Today, there is a growing recognition of the importance of local factors and community engagement in tourism development. Additionally, governments play a significant role in tourism governance, shaping regulatory frameworks and participating in decision-making processes at various levels. Therefore, comprehensively understanding the functioning of tourism requires a broader perspective on collaborative behaviour, which is critical for shaping outcomes and sustaining growth.

## The associational tourism economy

The associational nature of the way tourism functions has been implicitly acknowledged for a long time in tourism research and practice. The tourism economy embodies various business activities that co-create value to organise tourist experiences (Mwesiumo & Halpern, 2019). The associational tourism economy consists of associative relationships among businesses and organisations, including clusters (Cooke & Morgan, 1998; Möller & Halinen, 2017; Rosenfeld, 2005), networks (Camisón et al., 2017; Novelli et al., 2006; Shaw & Williams, 2009), and ecosystems (Araújo, 2022; Henche et al., 2020; Morgan et al., 2021; Sedarati et al., 2022; Vargas-Sánchez, 2019). Moreover, destinations themselves have been conceptualised as networks (Kofler et al., 2018), tourism clusters (Perkins et al., 2020), or business ecosystems (Steinbruch et al., 2022), leading research to focus on inter-organisation collaboration with an emphasis on geographical co-location (Nguyen et al., 2024). Although tourism networks, clusters, and business ecosystems are argued to be distinct concepts (Buhagiar, 2020; Wulf & Butel, 2017), the core mechanism within these concepts is the collaborative behaviour of businesses and organisations, no matter what associative framework it takes.

Around 80% of tourism businesses in the world are micro- to medium-sized (UNWTO, 2023), and therefore they have a limited capacity for growth that potentially could be improved through collaboration (Jesus & Franco, 2016). Collaborative behaviour is essential for the sustainable performance, survival, and innovation of tourism businesses (Hall & Williams, 2020). In this context, formal and informal networking helps to overcome resource limitation through associative behaviour among tourism businesses (Leick & Gretzinger, 2020; Zach, 2016; Zach & Racherla, 2011). Although business exchanges are vital for fostering collaborative behaviour among businesses, the characteristics specific to tourism businesses, such as their size, limited resources, and constrained knowledge, significantly influence the importance of social relationships in establishing and maintaining business relationships (Michael, 2007b; Novelli et al., 2006).

Associative behaviour is intrinsic to societies, driven by the fundamental nature of humans as social beings compelled to associate with one another (Curry & Dunbar, 2013; Granovetter, 1973, 1985). This behaviour enables businesses to enhance their competitive advantage and attract tourists through associative marketing, particularly amid growing tourist demand and increasing competition among destinations. The digitalisation of the economy further facilitates this by reducing the gap between consumers and producers (Buhalis & Leung, 2018), allowing for direct communications, and decreasing the need for intermediaries for marketing and distribution of tourist services and products. Collaborative behaviour has a significant connotation of survival, innovation, and resilience in tourism, which collectively influence the overall competitiveness of destinations (Fyall & Garrod, 2005; Marasco et al., 2018).

The concept of the associational tourism economy captures the economic relationships that arise from the tendency towards association among different businesses, groups, and/or organisations during the processes of developing, producing, marketing, distributing, and consuming tourism-related goods and services. The tourism economy is inherently associational because of the nature of tourist experiences as a production outcome of various businesses that emphasises the significance of collaboration. Collaboration intrinsically requires individuals to identify and associate themselves with other individuals through recognition of shared interests, background knowledge, and similarities, including belonging to a shared place, to reduce the uncertainty of collaboration (Curry & Dunbar, 2013). Tourism businesses produce their services associatively with other businesses and organisations, quasi-integrating production processes. Tourists also consume services associatively through hedonic connections with businesses, locations, experiences, and other tourists.

Therefore, globally tourism systems function associatively from business and consumer perspectives due to the tendency of individuals to associate with one another. The associational tourism economy addresses how and why tourist systems function, tourism businesses produce and market services and products, and tourists make decisions and consume these services and products. The term *associational tourism economy* refers to the dynamic economic activities of developing, producing, marketing, distributing, and consuming tourism goods and services driven by the need to belong to or be associated with groups, organisations, businesses, or communities. It signifies the individual contributions of various local businesses, organisations, and tourists in the functioning of global tourism systems.

The associational tourism economy highlights that economic activity extends beyond isolated business operations or networks of businesses, emphasising that business is conducted in a broader context of collaborative social dynamics (Frost & Crockett, 2007; Lazzeretti et al., 2019). As social beings, individuals naturally seek to belong within groups that engage in similar activities, share commonalities, and produce or consume similar products and services, reflecting the interplay between social and economic factors (Becattini, 1989, 1990). A shared place further fosters a sense of community, belongingness, and collaboration in a more effective way (Curry & Dunbar, 2013). The systemic nature of tourism reflects the complexity of the tourism systems that function interdependently. The associational tourism economy could be viewed from a quasi-integration perspective (Dietrich, 1994), suggesting a holistic view of tourism that acknowledges the partial integration of production processes involved in creating comprehensive tourism products. The associational tourism economy is the *sine qua non* of tourism systems' functioning.

## The factors driving collaborations

Today's business environment is becoming more unpredictable, beset by economic volatility, environmental challenges, and social issues. These issues are particularly pronounced in tourism, where many businesses are small- to micro-sized, making them especially susceptible to external pressures. In this climate, collaboration allows businesses to pool resources, share expertise, and align strategies to tackle the complex issues facing tourism development. This collective effort is vital for creating resilient, sustainable, and adaptive tourism systems that can navigate a rapidly changing landscape while maintaining the appeal and viability of destinations. In this context, collaboration is a vital response to the dynamic and unpredictable nature of the modern tourism business landscape. Thus, collaboration has become essential in the functioning of tourism systems, involving businesses, organisations, and tourists.

The associational tourism economy shifts the focus towards a broader collaborative domain, emphasising the complexities in addressing shared issues and goals. In tourism management and marketing, this collaborative domain is integral to developing effective strategies and solutions for complex system challenges. Collaboration enables businesses to enhance the overall tourist experience at destinations, which normally extends beyond the control of a single firm. Whether through self-organisation or with the assistance of destination management organisations, such collaborative relationships contribute to a more efficient tourism economic system. They achieve this by increasing the utilisation of resources, reducing transaction costs, and generating broader system-wide effects. In this way, cross-organisational collaboration becomes a foundational element of successful tourism management. Thus, associative behaviour within the tourism economy reflects the collaborative nature of relationships, aiming to address common interests, challenges, and conflicts in developing, producing, and marketing tourism products and services. Figure 1.1 demonstrates how collaboration types range from informal, issue-dependent to formal,

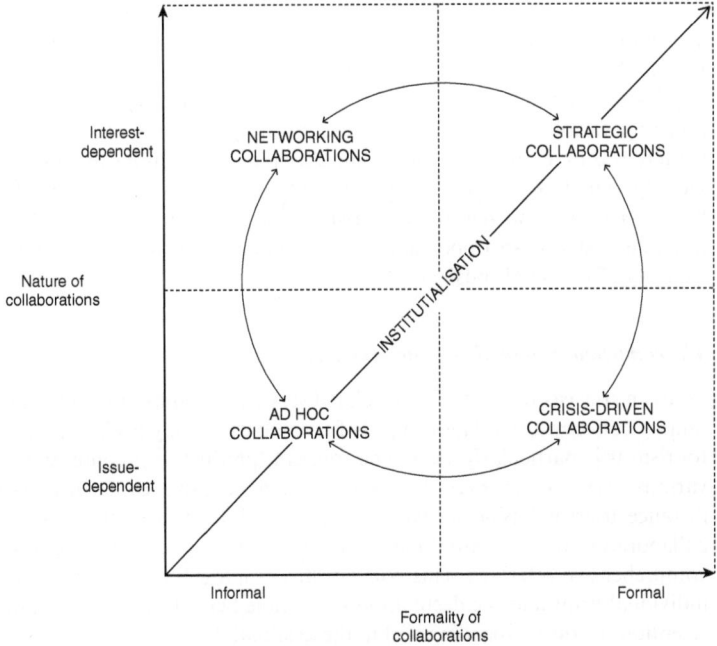

*Figure 1.1* Collaboration framework in tourism.

interest-dependent collaborations, with institutionalisation increasing from *ad hoc* to strategic approaches.

Two key forces underline the significance of collaboration in tourism: the inherent nature of tourism and the ongoing changes it is experiencing.

## Inherent characteristics of tourism

Tourism has been increasingly recognised as a key driver for regional development, particularly in rural and declining areas (Broegaard, 2020; Goulding et al., 2014; Helgadóttir & Dashper, 2021). The potential of tourism to stimulate local economies is attributed to several factors including its ability to attract visitors, create employment opportunities, and foster economic diversification. The relative simplicity of establishing tourism services allows small tourism businesses to leverage local resources and cultural assets to develop niche markets and enhance regional appeal, significantly contributing to regional development. However, the inherent characteristics of tourism require collaborations for tourism systems to function.

### *Complexity of tourism systems*

The complexity of tourism systems stems from the dispersion of production and consumption of tourist services across space and time. This implies that the impacts of tourism extend beyond the final destination, affecting various stages of the tourist journey, including travel to and from the destination. Numerous businesses and organisations provide services to tourists at different times and locations, often even before the trip begins. The systemic nature of tourism involves numerous complex and intricate systems, so opportunities and challenges extend beyond the capacity of any single business.

### *The compound nature of tourism products*

Tourism consists of a set of interrelated services. Establishing and maintaining collaborative relationships is fundamental to any business, but in tourism it is particularly vital. The tourism 'product' is a composite of various services perceived by tourists as a cohesive experience. For instance, tourism businesses can enhance their ability to attract tourists by collaborating with potential partners and bundle services creating more comprehensive offerings. This collaborative approach ensures that both individual firms and the destination as a whole benefit by promoting the retention of tourist spending within the destination.

*Variety of businesses serving tourists*

Many of businesses, while not directly associated with tourism, have tourists as part of their customer base. For example, hospitality businesses, such as accommodations, dining, and entertainment, are integral to the overall tourist experience. While often considered distinct, hospitality businesses are closely related to tourism. This wide range includes not only traditional tourism-related businesses but also retail establishments, transportation, and service providers that meet the needs of tourists. For example, shops, gas stations, and transport services may not be primarily focused on tourists but are essential for fulfilling their needs. The interconnectedness of these businesses highlights the multifaceted nature of tourism and underscores the importance of collaboration in tourism management and marketing to enhance the attractiveness of destinations and the tourist experience.

*Tourism business size*

The size of tourism businesses varies significantly, with a predominance of small- and micro-enterprises. While large companies play a crucial role in the tourism economy due to their substantial resources and market reach, micro- to medium-sized firms exhibit considerable potential for adaptation, growth, and innovation (Nordin, 2003; OECD, 2009). However, these tourism businesses often face constraints in resources such as knowledge, capital, and staffing which can restrict their ability to effectively address operational challenges. Collaboration with other businesses and organisations offers a strategic solution to mitigate the impact of these inherent resource limitations. As such, a collaborative approach not only addresses immediate operational challenges but also supports long-term growth and sustainability in tourism.

*Social nature of interactions in tourism*

The social nature of tourism is integral to its functioning, characterised by interactions among people and the provision of services by individuals for individuals. Tourism is fundamentally human-centred, with personal interactions playing a critical role in shaping experiences, delivering hospitality, and creating lasting memories. The social relationships among stakeholders are crucial for collaborations among businesses to develop cohesive and attractive tourism offerings. In essence, the ability to build strong relationships and maintain interactions is key to achieving a competitive advantage and ensuring the overall success of tourism destinations.

*Service features of tourism*

Tourism possesses distinct service features that differentiate it from physical products (see Kotler et al., 1996/2022). Unlike tangible goods, tourism services are produced and consumed simultaneously. Therefore, the quality and value of the service are experienced by tourists in real time, making the service delivery process critical to overall satisfaction. This simultaneous production and consumption introduce variability and perishability. The quality of the service can fluctuate based on numerous factors such as the timing of delivery, the interactions between service providers and tourists, and the experience context. Tourists travel to various destinations where tourism services are produced and delivered, and this mobility is a fundamental aspect of tourism. Service delivery is inherently tied to the location where it is provided, and the physical presence of tourists at the location is essential for the consumption of these services. As a result, understanding these service features is central to recognising the need for collaboration due to the complexities of tourism and the challenges and impacts associated with managing and delivering quality tourism services.

## Transformational changes in tourism

*Turbulent business environment*

The business environment is becoming increasingly turbulent due to a range of economic, environmental, and social issues. Tourist mobility exerts pressure on local communities and infrastructure. Economic instability driven by crises can disrupt international tourist flows and significantly impact destinations. Additionally, natural disasters and climate change can make destinations less attractive or even inaccessible. As these complexities unfold, the need for collaborative approaches becomes evident. No single tourism business or institution can effectively navigate the challenges posed by economic crises, natural disasters, and social tensions alone. Collaboration enables stakeholders, including governments, businesses, and local communities, to address the specific needs of a destination, fostering more resilient and sustainable tourism practices in the long term.

*Digitalisation*

The development of digital technologies has revolutionised tourism, transforming how tourists consume information and services, from booking tickets and accommodations to accessing destination information for decision-making. Digitalisation has also shifted the focus towards dissemination and knowledge acquisition for both businesses and organisations. While tourism in its early stages witnessed the emergence of tour operators facilitating connections between tourist services and packaging them, the rise of digital communication has shortened the distance between tourists and providers, diminishing the role of intermediaries. The digital transformation has reshaped interaction processes, offering new ways for engagement and convenience for both tourists and businesses. Information technologies have also created both new opportunities and challenges in how tourism businesses and destinations manage and market themselves.

### Increase in tourist numbers

Despite various disruptions slowing down economies, destinations worldwide have witnessed a significant increase in tourist numbers, consistently bouncing back to growth (UNWTO, 2023). This situation also presents both opportunities and challenges. Some regions, particularly rural and peripheral areas, face overdependence on international tourists, risking strain on local communities and their daily lives. Common issues include overtourism in major destinations, leading to overcrowding, shifts in resource consumption, and negative environmental impacts (Gössling & Peeters, 2015; Lew, 2014).

### Increasing competition

With the growth in global tourist numbers, competition among destinations has intensified. Urban areas experience disproportionate growth in tourism, while rural areas are frequently targeted by policies and research initiatives aimed at fostering economic prosperity or rejuvenating struggling areas (Goulding et al., 2014; Yang et al., 2021). Given the small scale of many tourism businesses, particularly in rural areas, independently enhancing a territory's appeal is becoming increasingly challenging without the collective involvement of tourism-related stakeholders. Increasing competition requires collaborative efforts to attract or redirect tourist flows and retain their interest. Thus, sustaining competitiveness requires ongoing collaboration among diverse stakeholders across various areas.

### Mass tourism to individualisation

Traditionally, tourism has been characterised by the standardised, packaged offerings often associated with mass tourism. However, there has been a significant shift from mass tourism to individualised experiences where many individual businesses create economies of scope, segment markets, and customise experiences (Poon, 1994). This change has emerged as a response to evolving tourist demands for authentic experiences and reflects a fundamental transformation in travel behaviour due to rising incomes, more affordable transportation, and easier access to information (Poon, 1993, 1994). It requires a re-evaluation of traditional tourism models as changing tourist behaviours significantly impact the management and marketing of businesses, organisations, and destinations. Consequently, the market has become more segmented, with diverse needs and preferences that demand flexibility and adaptability from destinations. The ability to offer diverse experiences is becoming a critical factor in destination competitiveness. This shift possesses opportunities and challenges that require collaboration among tourism-related stakeholders.

### *Anthropogenic pressure and environmental changes*

While mass tourism is attractive in terms of economies of scale, it appears unsustainable, especially given the increasing anthropogenic pressures associated with tourist mobility. Tourism contributes to carbon emissions (Lenzen et al., 2018); causes pollution and biodiversity loss; and pressures resource consumption such as water, energy, and land, slowly accumulating these impacts on destinations over time (Gössling & Peeters, 2015; Lew, 2014). Moreover, individual tourists may exert equal or greater impact on destinations, which often goes unnoticed compared to mass tourism's more easily observable effects. All of this requires the collaboration of different stakeholders to address emerging challenges and sustain environments.

### *Management is becoming marketing of destinations*

Marketing has emerged at the forefront of tourism development, with management increasingly led by marketing needs. Destination management differs significantly from traditional business management (Hall, 2005). Moreover, the aforementioned digitalisation contributes to the importance of maintaining a strong online presence and effectively marketing destinations for competitiveness. The emphasis shifts onto effective marketing of tourism offerings to outcompete other destinations. Consequently, destination management often prioritises marketing over managing the entire tourism systems. Moreover, DMOs are commonly referred to both as Destination Management Organisations and Destination Marketing Organisations (e.g., Jørgensen, 2017; Pike & Page, 2014; Quevedo et al., 2024), further blurring their functions. These institutions also often struggle to address local issues effectively. For instance, national or regional bodies may be geographically distant from the areas they oversee, making it difficult to fully comprehend the nuances of local needs and challenges. A lack of direct engagement with local communities can lead to solutions that may be well-intentioned but do not align with the actual needs of local areas. Integrating local context and insights into decision-making processes with a greater emphasis on collaboration, adaptability, and proactive problem-solving is essential for sustainable tourism.

Such complexity introduces governance challenges and requires a strategic collaborative approach to ensure the functionality of tourism systems. Effective collaboration among businesses, governmental organisations, educational institutions, and tourists is vital to address challenges related to resource utilisation, sustainability, and resilience. Collaborative behaviour is promising for mitigating the negative impacts of economic restructuring, especially in rural and peripheral areas (Hall, 2005). Tourism businesses can address individual challenges and unlock significant growth potential through collaborative efforts. Collaboration is essential for planning, development, management, and marketing in tourism, ensuring the success and competitiveness of tourism destinations and businesses. These efforts are crucial for overcoming weaknesses, building innovative capabilities, conducting effective marketing activities, and gaining sustainable competitive advantages (Sotiriadis et al., 2015). However, interest and conflict factors shape collaboration outcomes. Figure 1.2 categorises collaborations based on interest alignment and conflict nature.

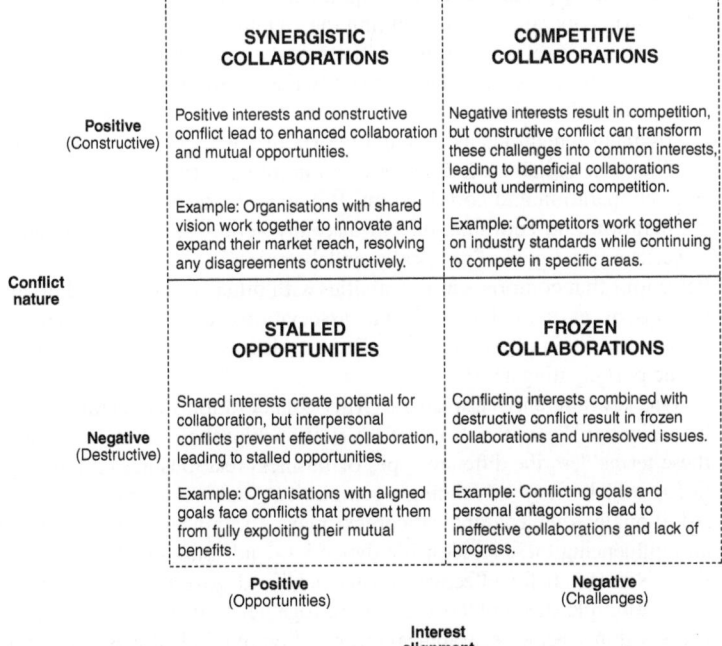

*Figure 1.2* Collaboration in tourism based on interest alignment and conflict nature.

**Collaboration and related concepts**

The terms 'collaboration', 'cooperation', and 'coordination' have become central themes in contemporary tourism research and policy. These concepts are often used together (Kofler et al., 2018) or interchangeably (Battisti & Peter, 2011) in the context of networking among stakeholders aimed at mutual benefits. However, caution is required when using these terms, as they have different meanings and implications for strategies and governance. For instance, tourism network policies designed to enhance networking have largely been replaced by tourism cluster policies with similar strategies (Perry, 2001, 2007). Despite their well-intentioned nature, many of these policies have failed (Markusen, 2003; Nishimura & Okamuro, 2011) or faded away (Ingley, 2008; Rodríguez et al., 2014). Therefore, a fundamental understanding of these terms and the approaches in use is significant for effective governance.

Collaboration is often perceived as a means of identifying common areas of interest among businesses and broader stakeholders. This process enables them to work together on activities such as, for example, lobbying, marketing, or promotion. Various tourism organisations serve as networks to bring parties together, emphasising the broader environment in which firms operate and highlighting the benefits of collaborative efforts. For instance, professional associations aim to address common issues faced by businesses with shared characteristics. Similarly, regional tourism organisations strive to build and market regional offerings. Networks are frequently regarded as a common form of collaborative organisation.

The term 'collaboration' serves as an umbrella term that encompasses both cooperation and coordination. It may be used to describe not only cooperation and coordination but also any action of working together towards mutual goals. Specifically, collaboration is often used to refer to behaviour that comprises joint activities with other businesses and organisations or participation in formal associations. In everyday language, these activities imply some form of networking assumed to offer benefits to the participating actors.

One key difference between the terms 'collaboration', 'cooperation', and 'coordination' lies in the nature of the relationships they reflect. In tourism, these terms describe different types of business relationships and interactions, which vary in timing, complexity, and relational dynamics. Additionally, the geographical context plays a significant role in shaping and influencing these relationship dynamics. Understanding these distinctions is essential for effective management and governance in tourism, where multiple stakeholders often work together to deliver tourist experiences within specific geographical settings. Figure 1.3 illustrates how 'coordination', 'cooperation', and 'collaboration' range from simple coordination for operational efficiency to cooperation and deeper collaboration in tourism, highlighting differences in timing, complexity, and relational dynamics among stakeholders.

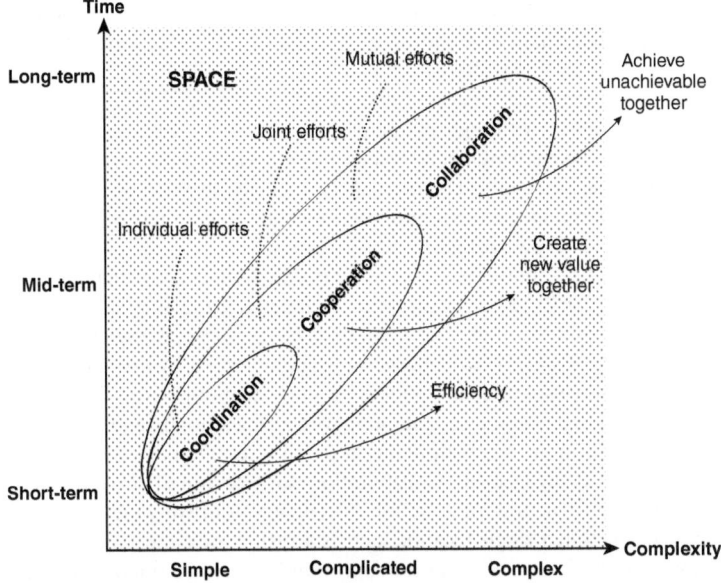

*Figure 1.3* Comparison of coordination, cooperation, and collaboration, emphasising variations in timing and complexity.

'Coordination' represents the simplest form of interaction among tourism stakeholders. It involves organising the *individual efforts* of different businesses or organisations to ensure operational *efficiency* without necessarily sharing resources or common strategic goals. For instance, a hotel coordinating with transport companies or local tour operators to schedule guest travel exemplifies coordination activity. The proximity of tourism businesses is an inherent location feature that significantly impacts how coordination occurs. In this context, coordination is typically short-term and *simple*, focusing on aligning activities to prevent overlap or conflict within a specific geographical area. While coordination facilitates smooth operations, it does not necessarily create new value or joint products. The tasks involved in coordination are manageable through careful analysis and proper planning.

'Cooperation' in tourism represents a complicated level of interaction. It involves *joint efforts*, where businesses work together to *create new value*, along with some degree of resource sharing such as marketing budgets, information, or expertise. For instance, several hotels within a specific region might cooperate with local attractions and restaurants to create a destination package that includes discounted stays, entry to attractions, and dining experiences. The geographical context plays a crucial role in shaping the nature of resource sharing and joint efforts.

Cooperation typically falls within a mid-term range and is more compli-cated than coordination, requiring a deeper level of interaction and com-mitment to achieving shared objectives. Unlike coordination, which focuses on operational efficiency, cooperation emphasises the creation of shared outcomes. Although cooperation involves multiple parties and tasks that can be challenging, it remains a *complicated* process rather than a complex one, as it can still be reasonably managed through *coordination*, structured planning, and execution, with geographical factors playing a significant role in this process.

'Collaboration', however, serves as an overarching concept that encom-passes both *coordination* and *cooperation*. It refers to relationships where multiple stakeholders come together to *achieve outcomes that would be unachievable individually*. Collaboration is characterised by *mutual efforts*, strategic resource sharing, and a focus on long-term objectives. For exam-ple, a tourism organisation collaborating with airlines, hotels, and local gov-ernments to promote a new destination on an international scale involves deep, long-term engagement. Collaboration can also aim to develop a des-tination's infrastructure, or establish sustainable tourism practices. This may occur within a single location, across a region, or even span multiple regions and nations. Collaboration is a *complex* form of interaction that often requires deep engagement and the integration of diverse perspectives with emergent behaviours among businesses and organisations arising from such interactions.

Unlike coordination and cooperation, *collaboration* inherently deals with the dimension of complexity, where uncertainty and emergent behav-iours necessitate more adaptive strategies. Effectively managing *collabora-tion* requires navigating a complex environment in which outcomes are not easily controlled or anticipated. The interconnected and interdepen-dent nature of the stakeholders involved, coupled with the influence of the geographical context, shapes these dynamics and demands flexibility and responsiveness in collaborative efforts.

These collaborations can be formal, such as structured agreements among businesses and organisations, or informal, such as following unwritten norms based on mutual interests. However, it is important to note that collaboration is not solely focused on long-term and strategic initiatives. Given that it incorporates elements of both cooperation and coordination, collaboration can also describe short- and mid-term behav-iours. The geographical context plays a dual role in collaboration, as it creates opportunities while also presenting challenges. Collaboration may extend over a large geographical area, across various local areas or even across multiple regions. This geographical diversity adds layers of com-plexity, particularly when regulatory environments, climates, and cultural differences, including business cultures, vary significantly between areas. The geographical context and intrinsic features of locations are integral to these interactions, significantly influencing the degree of complexity involved and the strategies required for success.

When comparing these terms, it is also important to distinguish between 'complicated' and 'complex'. A behaviour is considered 'complicated' when it involves multiple interconnected elements that, while challenging to manage, can ultimately be understood and addressed through structured approaches. Conversely, a 'complex' behaviour is characterised by unpredictable and emergent behaviours arising from the interactions of many components, making it far more difficult to manage or predict and requiring adaptive strategies. Collaborations, due to the diversity of interactions involved and the often long-term orientation, characterised by uncertainty, emergent behaviours, and challenges, regularly fall into the category of complex behaviours. Such interactions may also encounter unpredictable and emergent challenges such as changes in government policy or shifts in tourist behaviour. By contrast, cooperation and coordination are more likely to be classified as complicated or simple behaviours, respectively. Table 1.1 presents a comparative framework, detailing key characteristics that differentiate coordination, cooperation, and collaboration.

The complexity of collaboration in different contexts can be better understood through the lens of game theory and its theoretical models. Axelrod (1984/2009) explores various scenarios where questions of gains and losses influence participants' decisions, revealing that participants rarely choose to share gains equally. Early game-theoretic models highlighted the challenges of collaborative dilemmas in simple pair games. However, Axelrod (1997) later developed group interaction models, which illustrate the complex nature of multiparty collaborative efforts. In these models, the large-scale effects of locally interacting agents are referred to as 'emergent properties' of the system. Emergent properties can often be surprising because the full consequences of even simple interactions are difficult to anticipate. While neoclassical economic models assume that rational agents can efficiently reallocate resources, these models fall short when agents employ adaptive rather than optimising strategies (Axelrod, 1997). Therefore, simulation becomes necessary in such cases as it allows for the exploration of how agents adapt their behaviour based on past experiences, influencing their future actions.

While collaboration is essential in tourism, it is important to recognise that collaboration is not a panacea (Gray, 1989; Huxham, 1996). Collaboration is one of the strategies that businesses and broader tourism-related stakeholders can pursue in various areas, including business development, planning, marketing, and management. Acknowledging this fact is crucial as there may be a temptation to collaborate in every situation, or to assume that businesses will readily collaborate with each other, or within existing governance frameworks. Although collaboration can provide significant benefits, it is not

*Table 1.1* Key characteristics of coordination, cooperation, and collaboration

| Characteristics | Coordination | Cooperation | Collaboration |
| --- | --- | --- | --- |
| Meaning | Individual efforts in arranging and aligning activities to avoid conflict or duplication | Joint efforts towards shared goals of creating new value with some degree of mutual benefit | Mutual efforts to achieve shared value unachievable alone, involving close interaction and commitment |
| Scope | Narrow focus, typically task-based or operational | Broader scope than coordination, involving mutual aid on goal | Broad and long-term focus with combined efforts to innovate or achieve a significant impact |
| Complexity | Simple | Complicated | Complex |
| Focus | Efficiency | Common benefits | Common value |
| Goals | More individual | Shared | Collective |
| Management | Separate oversight by each party | Some joint management for shared areas | Joint or integrated management |
| Dependence | Partial independence | Partial dependence | Interdependence |
| Alignment | Task-oriented | Goal-oriented | Vision-oriented |
| Interaction processes | Structured, defined, efficiency-driven processes to ensure activities align | Flexible, allows for some adjustments to meet shared goals with less interdependence | Continuous, iterative, emergent processes that evolve with collaboration |
| Resources | Limited sharing of resources; each party uses primarily its own | Some shared resources for mutual benefit | Extensive use of resources to achieve shared outcomes |
| Information and knowledge sharing | Limited to necessary information, structured and often task-focused shared on a need-to-know basis | Frequent information exchange to align efforts | Transparent, continuous, and fluid encouraging creative input and vulnerability |
| Outcome | Efficient task completion | Mutual benefit, through compromise and mutual effort towards shared outcomes | Innovation, creativity, co-created results, and shared ownership |

always the optimal strategy. Businesses must choose their behaviours based on specific circumstances. While businesses may choose to compete, there are many ways to collaborate and find mutually beneficial decisions, including compromise, which facilitates common ground by limiting the demands of all participants.

Figure 1.4 demonstrates the strategic behaviour choices that tourism businesses can make, and how these behaviours align with varying degrees of relational interdependence and pursuit of interests. In the context of tourism, as behaviour shifts from competing to collaborating and back, business interests and interdependence change. Acknowledging the importance of working together in tourism, the transition between approaches reflects a strategic choice that can significantly impact the long-term viability and competitiveness of tourism businesses within a destination.

Businesses may adopt three distinct business behaviours, namely compete, compromise, and collaborate. They describe the strategic orientation of tourism businesses, ranging from focusing on their individual goals to pursuing strategies that benefit all parties involved. As businesses engage

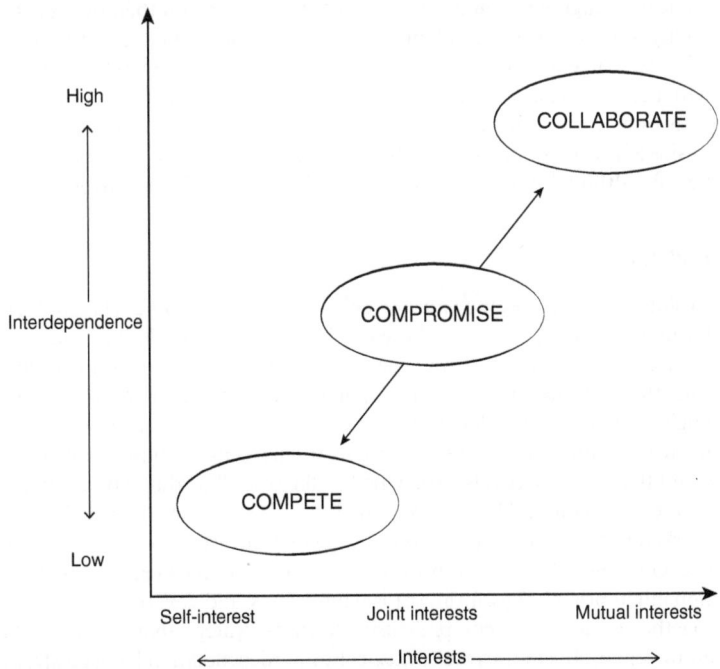

*Figure 1.4* Strategic tourism business behaviour.

in more compromise-oriented or collaborative strategies, the level of relational interdependence increases.

### Compete

Businesses operate largely independently, prioritising their own interests and maximising their own gains. For example, two neighbouring businesses might compete fiercely over price and amenities or implement aggressive marketing tactics to attract tourists, each trying to outperform the other and capture a larger share of the market, even if it means undercutting each other or other local competitors. Here, businesses operate more independently, with minimal collaboration, focusing on competitive behaviours with little regard for the broader context. Success is measured by individual gains rather than collective outcomes.

### Compromise

Businesses might acknowledge the need for some level of collaboration while still maintaining their individual interests. They start to see the value in engaging in compromises while still focusing on their own profitability. For instance, a hotel might agree to partner with another hotel to offer exchange of guests in case of overbooking, balancing its self-interest with the benefits of a collaborative arrangement. Such collaboration happens only if the terms are favourable and do not significantly reduce the business's own profit margins. Such compromises allow both parties to benefit, although they may not be fully aligned in their objectives.

### Collaborate

Businesses recognises the value of collaborations with other stakeholders for mutual benefits. They might work closely together to develop and promote comprehensive tourism experiences or establish broader norms and rules that enhance the destination's appeal. For example, a group of hotels might collaborate with local tour operators, restaurants, and cultural sites to create an all-inclusive travel package. In this scenario, businesses understand that their success is closely tied to the overall health and reputation of the destination. They actively work together, sharing knowledge and marketing efforts to create a more sustainable and attractive tourism offering. On a global scale, for example, a destination management organisation might bring hotels, airlines, local businesses, and government agencies together to develop a comprehensive tourism strategy that promotes the entire region. Businesses are strategically interdependent, relying on strong collaborations to succeed, which may lead to increased visitor numbers and shared economic gains, benefiting all parties involved.

However, collaboration does not oppose competition. To understand the interplay between collaboration and competition, it is essential to consider the concept of co-opetition (Chapter 3), when competitors simultaneously collaborate and compete. This perspective is rooted in the existence of common interests and shared geographical spaces. For instance, tourism businesses within a destination may collaborate to promote the area and enhance its attractiveness, benefiting all involved. They may share costs, engage in bulk procurement, and leverage collective bargaining power with suppliers while competing for tourists within the expanded market of the destination. However, as the market size grows and the number of businesses increases, finding common ground becomes increasingly challenging. Moreover, participants and new businesses will make rational choices based on self-interest strategies to maximize their profits.

Despite the potential benefits, there are several challenges that can hinder collaboration. Collaboration among tourism businesses is not guaranteed. Even when a common interest or issue exists (Gray, 1989), individual businesses may not perceive it as such – different actors will have different perceptions of a situation. Consequently, benefits will also be perceived differently by each party – what appears to be a gain for one party could be seen as negligible by another. Additionally, businesses may avoid engaging in collaborative efforts due to potential loss of independence, disagreements with others regarding their business model or standards, uncertainty about the benefits of collaboration, or concerns over power dynamics. Associations that represent common interests may also 'punish' new entrants or businesses that do not adhere to established agreements on norms and strategies by disengaging with such businesses. However, tourism businesses might be willing to pursue a new collaborative strategy if it demonstrates better results, indicating that collaboration, while complex, can be a viable course under the right conditions.

Collaboration also involves understanding different views of interests and conflicts to work together effectively. However, areas of conflict are often ignored, and when interests differ, collaboration is almost never pursued or even promoted. The desire to avoid conflict frequently leads to collaboration based solely on shared interests. While common interests facilitate collaboration, understanding and resolving conflicts can foster even deeper insights into problems and improve collaboration strategies. If the needs of all parties are unknown, it becomes impossible to determine the extent of collaboration, or to identify conflicts of interest. Each party often assumes that their perspective is obvious to everyone involved and that their approach is the only or correct way of doing things, even in positive collaborations. Thus, understanding participants' needs and perspectives is crucial for uncovering new opportunities.

Collaboration requires some form of contribution by an actor to a network organisation (Huxham, 1996). However, despite the perception of tourism organisations as inherently collaborative, not all organisations facilitate collaborative activities among participants. Instead, such network organisations, including government tourism organisations, often serve primarily as avenues for accessing knowledge and information, rather than fostering collaborative efforts. For instance, collaboration within these organisations frequently does not extend beyond marketing initiatives. Tourism businesses often pay fees to gain access to promotional opportunities through these organisations, yet they may not receive any tangible benefits such as an increase in customers. Furthermore, collaboration within these associations can often culminate in majority decisions on the best course of action that may lead to unsatisfactory outcomes for individual businesses. Businesses may choose to disengage or abstain from participation in these organisations if they disagree with the chosen development course. Consequently, the activities of such organisations tend to lean more towards coordinating members' behaviour rather than collaborating on a shared vision. Therefore, collaboration as a strategy involves navigating the complexities of cooperation and coordination in various strategic behaviours. While it can offer significant advantages, it also requires careful consideration of the specific context and potential challenges that may arise.

## Associational frameworks: clusters, networks, and ecosystems

Researchers and practitioners have explored various associational frameworks that facilitate collaboration among tourism businesses. Tourism literature emphasises the significance of different forms of associations, such as tourism networks (Hall et al., 1997; Hall & Williams, 2020), tourism clusters (Nordin, 2003; Teixeira et al., 2020), and, more recently, tourism ecosystems (Buhalis & Leung, 2018; Ogulin et al., 2016; Pencarelli, 2020). Collaborative arrangements are particularly relevant to tourism as the socio-economic features of networks correspond to the flows of goods, services, and tourists (Hall et al., 1997). The pioneering work on collaboration by Gray (1985) discussed the conditions significant for successful inter-organisational collaborations and stressed that the necessity of collaboration arises not only during a crisis. Collaborations are required when a problem is larger than what a single organisation can address, when traditional competing methods cannot resolve an issue, or when market turbulence increases (Gray, 1985).

Associational behaviour is represented by different forms of collaboration, varying from informal relationships to rigid contract arrangements (Hall et al., 1997; Selin & Chavez, 1995). Over the past decades, business competitiveness and the development of regional economies have been associated with the existence and growth of clusters (Cortright, 2006; Martínez-Pérez et al., 2019; Porter, 2000; Shakya, 2009; Simmie,

2004). It is argued that business clusters help overcome market challenges, increase the efficiency of resource usage, and enhance the attractiveness of a territory where tourism businesses are located. Tourism activities are characterised by the diversity of small firms and concentrations around tourist attractions or flows. Physical proximity positively enhances collaboration, increasing the frequency of interactions and making collaborations more likely, as there is some pre-existing interdependence between firms (Gray, 1985, 1989). Additionally, it is argued that knowledge generation and diffusion are spatially facilitated (Cooke & Morgan, 1998). Moreover, in tourism literature and policy-making, the importance of geographical proximity is also emphasised as it facilitates interactions, networking, and collaboration (Gnyawali & Srivastava, 2013; Quaranta et al., 2016; Rodríguez et al., 2014; Weidenfeld et al., 2011). In the context of tourism, the concepts of clusters and networks are closely related and have recently been further interlinked with the ecosystem approach (Araújo, 2022; Grumadaite, 2020; Henche et al., 2020; Morgan et al., 2021; Rachão et al., 2020; Sedarati et al., 2022; Steinbruch et al., 2022; Vargas-Sánchez, 2019). Overall, the focus within these concepts has shifted towards collaborative networking among tourism stakeholders (Camisón et al., 2017; Erkuş-Öztürk, 2009; Henche et al., 2020; Morgan et al., 2021; Sedarati et al., 2022; Shaw & Williams, 2009; Steinbruch et al., 2022; Vargas-Sánchez, 2019).

Clusters stem from agglomeration theories, which comprise a set of theories that explain why similar firms and industries tend to cluster in specific locations (Potter & Watts, 2010). The interest in the theory is attributed to the advantages that firms with similar or complementary activities can gain from being concentrated in a particular location due to agglomeration economies. Since tourism activities are often spatially fixed, co-location leads to the emergence of collaborative external economies that can result in reduced transaction costs, knowledge spillovers, innovation, and increased competitiveness (Marasco et al., 2018; Martínez-Pérez & Beauchesne, 2018). Agglomeration theory considers the potential of co-location for tourism businesses in the form of tourism clusters (Martínez-Pérez et al., 2019; Teixeira et al., 2020). Thus, geographical co-location provides the context for exploring collaborative behaviour due to (1) predetermined spatial context in tourism development and (2) collaborative behaviour advantages stemming from co-located tourism businesses. To clarify the concept of tourism clusters, they are defined as follows:

A tourism cluster is a geographically co-located concentration of businesses in a specific local area that offer similar or different services to an intertwining customer base, characterised by dynamic collaborative and competitive relationships rooted in interpersonal connections, shared interests, issues, and conflicts.

The primary argument for collaborative behaviour in tourism clusters is the numerous benefits that can arise from an ecosystem characterised by a shared vision, norms, values, and general 'rules of the game' (Aarikka-Stenroos & Ritala, 2017; Erkuş-Öztürk, 2011; Martínez-Pérez et al., 2021; Perkins et al., 2020; Rodríguez et al., 2014). In this context, the tourism cluster has become a recognisable type of ecosystem that facilitates collaborative behaviour in various forms. As complex adaptive systems (Grumadaite, 2020), business ecosystems demonstrate emergent properties (Anggraeni et al., 2007) that lead the systems to self-organise (Aarikka-Stenroos & Ritala, 2017). In relation to tourism, ecosystems can be characterised by the following definition:

> A tourism ecosystem is a complex adaptive system of tourism businesses, organisations, communities, tourists, and the natural environment that exhibits entropy, self-organisation, and emergent properties.

Networks are essential for understanding the interconnectedness of tourism businesses and their roles in facilitating collaboration. They are considered a source of sustainable development in tourism clusters (Lopes et al., 2019; Quaranta et al., 2016) and a key driver for ecosystems (Chen et al., 2022; Ness et al., 2024). Network theories consider processes connecting network characteristics to desired outcomes (Borgatti & Lopez-Kidwell, 2011). Networks have traditionally been seen as a way to coordinate business activities and drive innovation in tourism organisations (Hall, 2004; Hall et al., 1997). They provide a structure to create social capital for individuals (Burt, 1992) and communities (Putnam, 2000), offering an opportunity to explore why some groups achieve better results than others. In this context, tourism networks can be understood as follows:

> A tourism network refers to a geographically dispersed and interconnected system of tourism businesses, organisations, and individuals connected through mutually recognised formal or informal relationships to market products and services, address common issues, access resources and exchange information.

In the concepts of clusters, networks, and ecosystems within the associational tourism economy, collaboration plays a crucial role in enhancing the sustainability and competitiveness of tourism businesses. The interconnections among various stakeholders facilitate innovation and resilience in the face of market challenges. However, while the theoretical foundations of these collaborative frameworks provide valuable insights, their practical application often reveals inconsistencies and challenges that need to be acknowledged.

## Conclusion

The associational tourism economy presents both substantial opportunities and inherent challenges for collaboration among tourism businesses. The varying concepts of clusters, networks, and ecosystems highlight different aspects of collaborative frameworks, yet their interchangeable use can obscure their unique implications. Geographical proximity remains a pivotal factor in fostering collaboration, enhancing interactions, knowledge exchange, and competitive advantages among tourism businesses. However, the practical challenges of collaborative relationships are evident, particularly as many small businesses prioritise immediate operational needs over strategic collaborations. These challenges highlight the importance of understanding how to effectively cultivate collaborative environments within the framework of clusters, networks, and ecosystems. While collaboration offers significant benefits, it requires a nuanced approach tailored to the specific circumstances of each business. Acknowledging diverse interests, resolving conflicts, and promoting authentic collaboration are essential for effective tourism systems. These dynamics underscore the importance of understanding the geographical and relational aspects that drive successful collaborations in tourism. The next chapter will expand on the concepts of clusters, networks, and ecosystems, discussing their broader characteristics in tourism.

## References

Aarikka-Stenroos, L., & Ritala, P. (2017). Network management in the era of ecosystems: Systematic review and management framework. *Industrial Marketing Management, 67,* 23–36. 10.1016/j.indmarman.2017.08.010

Anggraeni, E., Den Hartigh, E., & Zegveld, M. (2007). Business ecosystem as a perspective for studying the relations between firms and their business networks. *ECCON 2007 Annual meeting, The Netherlands, Bergen aan Zee.*

Araújo, L. (2022). Measuring tourism success: How European national tourism organisations are shifting the paradigm. *Worldwide Hospitality and Tourism Themes, 14*(1), 79–84. 10.1108/WHATT-10-2021-0136

Axelrod, R. (1997). *The complexity of cooperation: Agent-based models of competition and collaboration.* Princeton University Press.

Axelrod, R. (2009). *The evolution of cooperation.* Basic Books. (Original work published 2009).

Battisti, M., & Peter, R. (2011). Compensating resource constraints? A discriminant analysis of collaborating and non-collaborating small businesses in New Zealand. *International Journal of Entrepreneurship and Small Business, 12*(2), 245–256. 10.1504/IJESB.2011.038539

Becattini, G. (1989). Sectors and/or districts: Some remarks on the conceptual foundations of industrial economics. In E. Goodman,

J. Bamford, & P. Saynor (Eds.), *Small firms and industrial districts in Italy* (pp. 123–135). Routledge.

Becattini, G. (1990). The Marshallian industrial district as a socio-economic notion. In F. Pyke, G. Becattini, & W. Sengenberger (Eds.), *Industrial districts and inter-firm co-operation in Italy* (pp. 37–51). International Institute for Labour Studies.

Borgatti, S., & Lopez-Kidwell, V. (2011). Network theory. In J. Scott & P. J. Carrington (Eds.), *The SAGE handbook of social network analysis.* Sage Publications.

Broegaard, R. B. (2020). Rural destination development contributions by outdoor tourism actors: A Bornholm case study. *Tourism Geographies.* 10.1080/14616688.2020.1795708

Buhagiar, K. (2020). Interorganizational learning in the tourism industry: Conceptualizing a multi-level typology. *The Learning Organization, 28*(2), 208–221. 10.1108/TLO-01-2020-0016

Buhalis, D., & Leung, R. (2018). Smart hospitality – Interconnectivity and interoperability towards an ecosystem. *International Journal of Hospitality Management, 71*, 41–50. 10.1016/j.ijhm.2017.11.011

Burt, R. S. (1992). *Structural holes: The social structure of competition.* Harvard University Press.

Camisón, C., Forés, B., & Boronat-Navarro, M. (2017). Cluster and firm-specific antecedents of organizational innovation. *Current Issues in Tourism, 20*(6), 617–646. 10.1080/13683500.2016.1177002

Chen, M.-K., Wu, S.-W., Huang, Y.-P., & Chang, F.-J. (2022). The key success factors for the operation of SME cluster business ecosystem. *Sustainability, 14*(14), 8236. 10.3390/su14148236

Cooke, P., & Morgan, K. (1993). The network paradigm: New departures in corporate and regional development. *Environment and Planning D: Society and Space, 11*(5), 543–564. 10.1068/d110543

Cooke, P., & Morgan, K. (1998). *The associational economy: Firms, regions, and innovation.* Oxford University Press.

Cortright, J. (2006). *Making sense of clusters: Regional competitiveness and economic development.* The Brookings Institution.

Curry, O., & Dunbar, R. I. M. (2013). Do birds of a feather flock together? *Human Nature, 24*(3), 336–347. 10.1007/s12110-013-9174-z

De Man, A. (2004). *The network economy: Strategy, structure and management.* Edward Elgar Publishing.

Dietrich, M. (1994). The economics of quasi-integration. *Review of Political Economy, 6*(1), 1–18. 10.1080/09538259400000001

Erkuş-Öztürk, H. (2009). The role of cluster types and firm size in designing the level of network relations: The experience of the Antalya tourism region. *Tourism Management, 30*(4), 589–597. 10.1016/j.tourman.2008.10.008

Erkuş-Öztürk, H. (2011). Emerging importance of institutional capacity for the growth of tourism clusters: The case of Antalya. *European Planning Studies, 19*(10), 1735–1753. 10.1080/09654313.2011.614384

Frost, M., & Crockett, J. (2007). Communities of practice, clusters or networks? Prospects for collaborative business arrangements in the mining and engineering sector, central western New South Wales. In P. Basu, G.

O'Neill, & A. Travaglione (Eds.), *Engagement & change: Exploring management, economic and finance fimplications of a globalising environment* (pp. 249–260). Australian Academic Press.

Fyall, A., & Garrod, B. (2005). *Tourism marketing: A collaborative approach.* Channel View Publications.

Gnyawali, D. R., & Srivastava, M. K. (2013). Complementary effects of clusters and networks on firm innovation: A conceptual model. *Journal of Engineering and Technology Management, 30*(1), 1–20. 10.1016/j.jengtecman.2012.11.001

Gössling, S., & Peeters, P. (2015). Assessing tourism's global environmental impact 1900-2050. *Journal of Sustainable Tourism, 23*(5), 639–659. 10.1080/09669582.2015.1008500

Goulding, R., Horan, E., & Tozzi, L. (2014). The importance of sustainable tourism in reversing the trend in the economic downturn and population decline of rural communities. *PASOS Revista De Turismo Y Patrimonio Cultural, 12*(3), 549–563. 10.25145/j.pasos.2014.12.041

Granovetter, M. (1973). The strength of weak ties. *The American Journal of Sociology, 78*(6), 1360–1380. 10.1086/225469

Granovetter, M. (1985). Economic action and social structure: The problem of embeddedness. *The American Journal of Sociology, 91*(3), 481–510. 10.1086/228311

Gray, B. (1985). Conditions facilitating interorganizational collaboration. *Human Relations, 38*(10), 911–936. 10.1177/001872678503801001

Gray, B. (1989). *Collaborating: Finding common ground for multiparty problems.* Jossey-Bass.

Grumadaite, K. (2020). Sustainable emergence of regional tourism clusters: The approach of complexity theory. *International Conference on Tourism Research.* 10.34190/IRT.20.038

Hall, C. M. (2004). Small firms and wine and food tourism in New Zealand: Issues of collaboration, clusters and lifestyles. In R. Thomas (Ed.), *Small firms in tourism* (pp. 167–181). Taylor & Francis.

Hall, C. M. (2005). *Tourism: Rethinking the social science of mobility.* Pearson Education.

Hall, C. M., & Michael, E. J. (2006). Issues in regional development. In E. J. Michael (Ed.), *Micro-clusters and networks: The growth of tourism* (pp. 27–40). Routledge.

Hall, C. M., & Williams, A. M. (2020). *Tourism and innovation.* Routledge.

Hall, C.M., Cambourne, B., Macionis, N., & Johnson, G. (1997). Wine tourism and network development in Australia and New Zealand: Review, establishment and prospects. *International Journal of Wine Marketing, 9*(2), 5–31. 10.1108/eb008668

Helgadóttir, G., & Dashper, K. (2021). 20 years of Nordic rural tourism research: A review and future research agenda. *Scandinavian Journal of Hospitality and Tourism, 21*(1), 60–69. 10.1080/15022250.2020.1823246

Henche, B. G., Salvaj, E., & Cuesta-Valiño, P. (2020). A sustainable management model for cultural creative tourism ecosystems. *Sustainability, 12*(22), 1–21. 10.3390/su12229554

Huxham, C. (Ed.). (1996). *Creating collaborative advantage.* SAGE Publications.

Ingley, C. (2008). *The cluster concept: Cooperative networks and replicability*. Paper presented at the *Conference 'International Council for Small Businesses'*. Naples, Italy.

Jesus, C., & Franco, M. (2016). Cooperation networks in tourism: A study of hotels and rural tourism establishments in an inland region of Portugal. *Journal of Hospitality and Tourism Management, 29*, 165–175. 10.1016/j.jhtm.2016.07.005

Johansson, B., Karlsson, C., & Westin, L. (Eds.). (2012). *Patterns of a network economy*. Springer.

Jørgensen, M. T. (2017). Developing a holistic framework for analysis of destination management and/or marketing organizations: Six Danish destinations. *Journal of Travel & Tourism Marketing, 34*(5), 624–635. 10.1080/10548408.2016.1209152

Kim, N., & Shim, C. (2018). Social capital, knowledge sharing and innovation of small- and medium-sized enterprises in a tourism cluster. *International Journal of Contemporary Hospitality Management, 30*(6), 2417–2437. 10.1108/ijchm-07-2016-0392

Kofler, I., Marcher, A., Volgger, M., & Pechlaner, H. (2018). The special characteristics of tourism innovation networks: The case of the regional innovation system in South Tyrol. *Journal of Hospitality and Tourism Management, 37*, 68–75. 10.1016/j.jhtm.2018.09.004

Kotler, P., Bowen, J. T., Baloglu, S., & Morosan, C. (2022). *Marketing for hospitality and tourism* (8th global ed.). Pearson. (Original work published 1996)

Lazzeretti, L., Capone, F., Caloffi, A., & Sedita, S. R. (2019). Rethinking clusters. Towards a new research agenda for cluster research. *European Planning Studies, 27*(10), 1879–1903. 10.1080/09654313.2019.1650899

Leick, B., & Gretzinger, S. (2020). Business networking in organisationally thin regions: A case study on network brokers, SMEs and knowledge-sharing. *Journal of Small Business and Enterprise Development, 27*(5), 839–861. 10.1108/JSBED-12-2019-0393

Lenzen, M., Sun, Y.-Y., Faturay, F., Ting, Y.-P., Geschke, A., & Malik, A. (2018). The carbon footprint of global tourism. *Nature Climate Change, 8*(6), 522–528. 10.1038/s41558-018-0141-x

Lew, A. A. (2014). Scale, change and resilience in community tourism planning. *Tourism Geographies, 16*(1), 14–22. 10.1080/14616688.2013.864325

Lopes, H. D. S., Remoaldo, P., & Ribeiro, V. (2019). Residents' perceptions of tourism activity in a rural North-Eastern Portuguese community: A cluster analysis. *Bulletin of Geography, 46*(46), 119–135. 10.2478/bog-2019-0038

Marasco, A., De Martino, M., Magnotti, F., & Morvillo, A. (2018). Collaborative innovation in tourism and hospitality: A systematic review of the literature. *International Journal of Contemporary Hospitality Management, 30*(6), 2364–2395. 10.1108/ijchm-01-2018-0043

Markusen, A. (2003). Fuzzy concepts, scanty evidence, policy distance: The case for rigour and policy relevance in critical regional studies. *Regional Studies, 37*(6–7), 701–717. 10.1080/0034340032000108796

Martínez-Pérez, Á., & Beauchesne, M.-M. (2018). Overcoming the dark side of closed networks in cultural tourism clusters: The importance of diverse networks. *Cornell Hospitality Quarterly*, *59*(3), 239–256. 10.1177/1938965517734938

Martínez-Pérez, Á., Elche, D., García-Villaverde, P. M., & Parra-Requena, G. (2019). Cultural tourism clusters: Social capital, relations with institutions, and radical innovation. *Journal of Travel Research*, *58*(5), 793–807. 10.1177/0047287518778147

Martínez-Pérez, Á., Elche, D., & García-Villaverde, P. M. (2021). Bridging capital and performance in clustered firms: The heterogeneous effect of knowledge strategy. *Tourism Management*, *85*. 10.1016/j. tourman.2020.104264

Michael, E. J. (2007a). *Micro-clusters and networks: The growth of tourism*. Routledge. 10.4324/9780080464909

Michael, E. J. (2007b). Micro-clusters in tourism. In E. J. Michael (Ed.), *Micro-clusters and networks: The growth of tourism* (pp. 33–42). Routledge.

Molenaar, C. (2020). *The end of competition: The impact of the network economy*. World Scientific Publishing. 10.1142/11608

Möller, K., & Halinen, A. (2017). Managing business and innovation networks – From strategic nets to business fields and ecosystems. *Industrial Marketing Management*, *67*, 5–22. 10.1016/j.indmarman.2017.09.018

Morgan, M. O., Okon, E. E., Emu, W. H., Olubomi, O. I. E., & Edodi, H. U. (2021). Tourism management: A panacea for sustainability of hospitality industry. *Geojournal of Tourism and Geosites*, *37*(3), 783–791. 10.30892/GTG.37307-709

Mwesiumo, D., & Halpern, N. (2019). A review of empirical research on interorganizational relations in tourism. *Current Issues in Tourism*, *22*(4), 428–455. 10.1080/13683500.2017.1390554

Ness, H., Aarstad, J., & Haugland, S. A. (2024). Structural networks and dyadic negotiations in tourism destination ecosystems. *International Journal of Contemporary Hospitality Management*, *36*(2), 379–399. 10.1108/IJCHM-03-2022-0309

Nguyen, T. Q. T., Nguyen, V. T., Hoang, T. T. H., Tran, T. H. T., & Nguyen, T. P. T. (2024). Social networking, environmental awareness and sustainable tourism development in Da Nang, Vietnam. *Tourism and Hospitality Research*. 10.1177/14673584241234269

Nishimura, J., & Okamuro, H. (2011). Subsidy and networking: The effects of direct and indirect support programs of the cluster policy. *Research Policy*, *40*(5), 714–727. 10.1016/j.respol.2011.01.011

Nordin, S. (2003). *Tourism clustering & innovation: Paths to economic growth & development*. Etour.

Novelli, M., Schmitz, B., & Spencer, T. (2006). Networks, clusters and innovation in tourism: A UK experience. *Tourism Management*, *27*(6), 1141–1152. 10.1016/j.tourman.2005.11.011

OECD. (2009). *Clusters, innovation and entrepreneurship*. 10.1787/978926 4044326-en

Ogulin, R., Selen, W., & Houghton, L. (2016). Coordination in a tourism ecosystem: Methods to tackle wicked problems. *Emergence: Complexity and Organization*, *18*(1). 10.emerg/10.17357.1f1e70d186bad562e656d3e 1d25c3887

Pencarelli, T. (2020). The digital revolution in the travel and tourism industry. *Information Technology & Tourism*, *22*(3), 455–476. 10.1007/ s40558-019-00160-3

Perkins, R., Khoo-Lattimore, C., & Arcodia, C. (2020). Understanding the contribution of stakeholder collaboration towards regional destination branding: A systematic narrative literature review. *Journal of Hospitality and Tourism Management*, *43*, 250–258. 10.1016/j. jhtm.2020.04.008

Perry, M. (2001). *Shared trust in New Zealand: Strategies for a small industrial country*. Institute of Policy Studies, Victoria University of Wellington.

Perry, M. (2007). From networks to clusters and back again: A decade of unsatisfied policy aspiration in New Zealand. In R. MacGregor & A. Hodgkinson (Eds.), *Small business clustering technologies: Applications in marketing, management, IT and economics* (pp. 160–183). IGI Global.

Phillipson, J., Gorton, M., & Laschewski, L. (2006). Local business co-operation and the dilemmas of collective action: Rural micro-business networks in the North of England. *Sociologia Ruralis*, *46*(1), 40–60. 10.1111/j.1467-9523.2006.00401.x

Pike, S., & Page, S. J. (2014). Destination marketing organizations and destination marketing: A narrative analysis of the literature. *Tourism Management*, *41*, 202–227. 10.1016/j.tourman.2013.09.009

Poon, A. (1993). *Tourism, technology and competitive strategies*. CAB international.

Poon, A. (1994). The 'new tourism' revolution. *Tourism Management*, *15*(2), 91–92. 10.1016/0261-5177(94)90001-9

Porter, M. E. (2000). Location, competition, and economic development: Local clusters in a global economy. *Economic Development Quarterly*, *14*(1), 15–34. 10.1177/089124240001400105

Potter, A., & Watts, H. D. (2010). Evolutionary agglomeration theory: Increasing returns, diminishing returns, and the industry life cycle. *Journal of Economic Geography*, *11*(3), 417–455. 10.1093/jeg/lbq004

Powell, W. (1991). Neither market nor hierarchy: Network forms of organization. In G. Thomson, J. Frances, R. Levačić, & J. Mitchell (Eds.), *Markets, hierarchies, networks – The coordination of social life* (pp. 265–276). SAGE Publications.

Putnam, R. (2000). *Bowling alone: The collapse and revival of American community*. Simon & Schuster.

Quaranta, G., Citro, E., & Salvia, R. (2016). Economic and social sustainable synergies to promote innovations in rural tourism and local development. *Sustainability*, *8*(7), 668. 10.3390/su8070668

Quevedo, L., Herrera, R., Aldaz, S., Godoy, S., & Merino, K. (2024). Stakeholders' perceptions about a destination management organization (DMO): A case study in chimborazo, Ecuador. *Journal of Educational and Social Research*, *14*(1), 234–242. 10.36941/jesr-2024-0020

Rachão, S., Breda, Z., Fernandes, C., & Joukes, V. (2020). Cocreation of tourism experiences: Are food-related activities being explored? *British Food Journal, 122*(3), 910–928. 10.1108/BFJ-10-2019-0769

Rodríguez, I., Williams, A. M., & Hall, C. M. (2014). Tourism innovation policy: Implementation and outcomes. *Annals of Tourism Research, 49*, 76–93. 10.1016/j.annals.2014.08.004

Rosenfeld, S. A. (2005). Industry clusters: Business choice, policy outcome, or branding strategy? *Journal of New Business Ideas and Trends, 3*(2), 4–13.

Sedarati, P., Serra, F. M. D., & Jakulin, T. J. (2022). Systems approach to model smart tourism ecosystems. *International Journal for Quality Research, 16*(1), 285–306. 10.24874/IJQR16.01-20

Selin, S., & Chavez, D. (1995). Developing an evolutionary tourism partnership model. *Annals of Tourism Research, 22*(4), 844–856. 10.1016/0160-7383(95)00017-x

Shakya, M. (2009). Clusters for competitiveness: A practical guide and policy implications for developing cluster initiatives. *SSRN Electronic Journal.* 10.2139/ssrn.1392479

Shaw, G., & Williams, A. (2009). Knowledge transfer and management in tourism organisations: An emerging research agenda. *Tourism Management (1982), 30*(3), 325–335. 10.1016/j.tourman.2008.02.023

Simmie, J. (2004). Innovation and clustering in the globalised international economy. *Urban Studies, 41*(5/6), 1095–1112. 10.1080/00420980410001675823

Sorbe, S., Gal, P., & Millot, V. (2018). Can productivity still grow in service-based economies? Literature overview and preliminary evidence from OECD countries. *OECD Economics Department Working Papers*, 1531. 10.1787/4458ec7b-en

Sotiriadis, M., Gursoy, D., & Saayman, M. (2015). Conclusions: Issues and challenges for collaborative forms in tourism business and destinations. In D. Gursoy, M. Saayman, & M. Sotiriadis (Eds.), *Collaboration in tourism businesses and destinations: A handbook* (pp. 321–330). Emerald.

Steinbruch, F. K., Nascimento, L. d. S., & de Menezes, D. C. (2022). The role of trust in innovation ecosystems. *Journal of Business & Industrial Marketing, 37*(1), 195–208. 10.1108/JBIM-08-2020-0395

Streeck, W., & Schmitter, P. C. (1991). Community, market, state – and associations? The prospective contribution of interest governance to social order. In G. Thomson, J. Frances, R. Levačić, & J. Mitchell (Eds.), *Markets, hierarchies, networks – The coordination of socail life* (pp. 227–239). SAGE Publications.

Teixeira, S. J., João, J. M. F., & Correia, R. C. (2020). What do we know about tourism cluster and insular economy: A bibliometric study. *Journal of Spatial and Organizational Dynamics, 8*(2), 107–128.

UNWTO. (2013). *Sustainable tourism for development guidebook – Enhancing capacities for sustainable tourism for development in developing countries.* UNWTO. 10.18111/9789284415496

UNWTO. (2023). *International tourism highlights, 2023 Edition – The impact of COVID-19 on tourism (2020–2022).* 10.18111/9789284424986

Vargas-Sánchez, A. (2019). The new face of the tourism industry under a circular economy. *Journal of Tourism Futures*, 7(2), 203–208. 10.1108/ JTF-08-2019-0077

Weidenfeld, A., Butler, R., & Williams, A. W. (2011). The role of clustering, cooperation and complementarities in the visitor attraction sector. *Current Issues in Tourism*, 14(7), 595–629. 10.1080/13683500.2010.517312

Wulf, A., & Butel, L. (2017). Knowledge sharing and collaborative relationships in business ecosystems and networks: A definition and a demarcation. *Industrial Management & Data Systems*, 117(7), 1407–1425. 10.1108/IMDS-09-2016-0408

Yang, J., Yang, R., Chen, M.-H., Su, C.-H., Zhi, Y., & Xi, J. (2021). Effects of rural revitalization on rural tourism. *Journal of Hospitality and Tourism Management*, 47, 35–45. 10.1016/j.jhtm.2021.02.008

Zach, F. (2016). Collaboration for innovation in tourism organizations: Leadership support, innovation formality, and communication. *Journal of Hospitality and Tourism Research*, 40(3), 271–290. 10.1177/109634801 3495694

Zach, F., & Racherla, P. (2011). Assessing the value of collaborations in tourism networks: A case study of Elkhart County, Indiana. *Journal of Travel & Tourism Marketing*, 28(1), 97–110. 10.1080/10548408.2011. 535446

# 2 The emergence of clusters, networks, and ecosystems

## Introduction

Regional economists have often neglected the significance of tourism for regional economic development, even though tourism is becoming central to the economic policies of many regions (Calero & Turner, 2020). In the light of disproportional economic development between urban and rural areas (Safonov & Hall, 2023), tourism clusters are frequently seen as a tool to stimulate economic opportunities in rural and peripheral communities, which often struggle to compete with the concentrated capital in urban centres (Michael, 2007). Leveraging cultural heritage, authentic traditions, and local uniqueness means such areas have the potential to develop tourism offerings that contribute to local economies (Tolstad, 2014). However, small tourism businesses must collaborate to create such offerings and achieve sustainable outcomes. Given that most tourism businesses represent micro to medium businesses (UNWTO, 2023), the associational conceptualising of tourism development is particularly significant.

Collaboration among tourism businesses, organisations, and communities has been viewed through the related concepts of tourism 'clusters', alongside 'networks' (Halme, 2001; Hemphälä & Magnusson, 2012; Novelli et al., 2006; Perkins et al., 2022) and 'business ecosystems' (Buhalis & Leung, 2018; Ogulin et al., 2016; Pencarelli, 2020; Steinbruch et al., 2022). The concept of business ecosystems has only recently gained prominence in tourism studies (Steinbruch et al., 2022), while networks have long been acknowledged as key components of collaborative behaviour in tourism (Fyall & Garrod, 2005). Geographical features are inherently tied to the organisation, functioning, and development of tourism, making the cluster framework particularly relevant. The spatial concentration of businesses around tourist attractions or flows driven by natural or climatic resources makes clusters an attractive framework for tourism-related regional development. The cluster concept remains widely accepted as representing the spatial manifestation of tourism, capable of fostering

DOI: 10.4324/9781003293606-2

innovation (Nordin, 2003; Weidenfeld & Hall, 2014) and promoting sustainable outcomes (Martínez-Pérez et al., 2019; Tolstad, 2014).

While these concepts may seem distinct, they are closely intertwined in practice and in the literature, as all emphasise networking opportunities as the foundation for efficient and sustainable tourism development. However, research often fails to clearly define the distinctions between them. For example, the tourism ecosystem approach is often used without proper definition (Araújo, 2022; Morgan et al., 2021; Philipp et al., 2022; Rachão et al., 2020), interpreting it in the same way as existing network and cluster concepts (Grumadaite, 2020; Henche et al., 2020; Madanaguli et al., 2022; Sedarati et al., 2022). Many studies use these terms without fully clarifying their specific meanings or differences, often applying one concept to explain another (e.g., Wulf & Butel, 2017). Tourism research recognises the interconnectedness of these concepts, even though they have evolved separately. Despite this, the literature consistently highlights shared features, such as knowledge exchange, innovation, social capital, and a collaborative atmosphere, all of which are forged through frequent interactions among businesses, organisations, and communities. This chapter begins with a review of the cluster concept and then examines the concepts of networks and ecosystems as a broader state of tourism systems. Although these terms are sometimes used interchangeably, leading to confusion and blurred boundaries, they serve as a framework within which to explore the interdependencies among tourism stakeholders.

## Clusters

The cluster concept has been an essential topic in economic development for several decades (Chandrashekar & Mungila Hillemane, 2018; Fafurida & Mulyaningsih, 2023; Lindqvist et al., 2013; Sölvell, 2009). Since tourism businesses are often spatially fixed and tend to co-locate naturally within a destination context (Hall & Williams, 2020), the discussion of collaboration in tourism is relevant to the cluster concept. The industrial cluster concept serves as the foundation for the conceptualisation of the tourism cluster. It is considered one of the significant drivers for collaborations between businesses at destinations and regional development agencies, representing an advanced framework in the associational economy (Rosenfeld, 2005).

Interest in clusters among researchers, economists, and policymakers arose in the 1990s when Porter (1990) introduced the concept as part of his theory of national competitive advantage. While the cluster concept is often linked to Porter, it is essential to note that his emphasis was on the theory of competitiveness, in which clusters are a part of his framework related to the 'clustering of competitive industries' (Porter, 1990, p.148).

With promising prospects for countries and regions to improve their international competitiveness, the cluster concept has diffused widely through economic and regional development policy-making. It became the leading strategy for economic development among international organisations such as the OECD (2005, 2009). Since the publication of Porter's book in the 1990s, numerous attempts have been made to interpret and explain the functioning of clusters in various countries (e.g., Gilding et al., 2020; Maté-Sánchez-Val & Harris, 2018; Ning, 2021; Novotná & Novotný, 2019). The results of cluster functioning are still inconsistent, varying from positive consequences (Lai et al., 2014; Morosini, 2004) to only a small impact on economic development (Bakarić, 2017; Gilding et al., 2020). Despite this variability, the literature still frequently references Porter (1990) when defining a cluster (e.g., Gohr & Oliveira, 2019; Lechner & Leyronas, 2012; Nie et al., 2020).

In attempts to discover new facets of clusters, the literature often takes a socio-economic perspective grounded in the principle of the social embeddedness of economic activity. The network approach has become dominant in cluster literature (Cruz & Teixeira, 2010; Lazzeretti et al., 2014). However, this has led to the blending of different concepts, with 'clusters', 'networks', and 'ecosystems' being used interchangeably, often without proper differentiation. While these approaches intersect, they are not well-synthesised together. This creates uncertainty and confusion regarding their understanding and application. For example, definitions of clusters vary widely, ranging from simple concentration to informal networking of businesses (Swann, 2009). However, practical implementation is often limited to the formation of associations, as the initial stage of cluster development typically involves establishing an organisation for funding and support purposes. Policies also aim to boost networking and joint activities through different types of associative forms. For instance, network policies have been replaced by cluster policies with the same focus on collaboration (Perry, 2007). Nevertheless, tourism presents an attractive area for applying the cluster concept, as it is characterised by business concentration and the importance of collective actions (Havierniková et al., 2017; Sigurðardóttir & Steinthorsson, 2018).

### The industrial cluster

The clustering of industries is usually explained through Porter's theory of competitiveness. Porter (1990) asserts that competitive industries 'are not spread evenly through the economy but are connected in … clusters consisting of industries related by links of various kinds' (Porter, 1990, p. 131). In his view, clusters of several industries are the gravity centres for economic development. Industries utilise 'common inputs, skills and infrastructure' that 'stimulates government bodies, educational institutions, firms, and individuals to invest in relevant factor creation or factor-creating mechanisms' (Porter, 1990, p. 135). However, Porter (1990) did not explicitly define clusters in his original work. In the introduction added to the original book in 1998, he claims to have introduced 'the concept of clusters, or groups of interconnected firms, suppliers, related industries, and specialized institutions in particular fields that are present in particular locations' (Porter, 1998a/1990, p. xxii).

While the definitions of clusters vary throughout Porter's works (Porter, 1998b, pp. 102, 109, 113, Porter, 2000, p. 32), it is evident that these definitions share some similarities, particularly the emphasis on the presence of different interconnected organisations in a close geographical location. However, they are not consistent in determining what a cluster is, varying from a group of firms (Porter, 1998b, p. 102) to a network (Porter, 1998b, p. 113) and a concentration of skills and knowledge (Porter, 2000, p. 32). While the general idea appears clear, the level of analysis has shifted from a grouping of industries in Porter (1990) to a concentration of firms in Porter (1998b), complicating the conceptualisation.

Interestingly, the cluster definition itself may not be essential to Porter. He states that 'appropriate definition of a cluster can differ in different locations, depending on the segments in which the member companies compete and the strategies they employ' (Porter, 1998b, p. 105). Thus, defining a cluster is left to the researcher, practitioner, or policymaker within the general context of Porter's proposition. A notable example of this approach is the OECD (2005), where various countries employed different notions for their cluster programmes under the umbrella of the OECD's Local Employment and Economic Development (LEED) Programme. Overall, most definitions imply interconnectedness, cooperation, and the collective efforts of firms in close geographical proximity.

The cluster concept was originally proposed as an analytical framework, but it is often misapplied (Sölvell, 2009, p. 125). Sölvell (2009) argues that policymakers tend to undervalue the role of competition in favour of collaboration, relying on planning while overlooking the fact that the cluster concept was not intended as a policy initiative or programme. Even with the confusion in results (Vom Hofe & Chen, 2006) and reasonable critiques (Martin & Sunley, 2003; Maskell & Malmberg, 2002), the cluster concept, initially part of the competitiveness theory in Porter (1990), has evolved into what is now referred to as 'the theory of clusters' in Porter (1998b).

### Emerging concerns regarding the industrial cluster concept

The cluster concept proved to be timely as it became widely applied by governments. However, Porter, trying to broaden agglomeration effects to international competitiveness, combines different theories in the manifestation of clusters. Porter's representation of clusters may confuse as it represents different standpoints.

### Terminology and definition

The thread of arguments becomes obscured by the overuse of various terms and definitions. For instance, terms such as 'domestic cluster' (e.g., Porter, 1990, p. 636), 'local cluster' (e.g., Porter, 1990, p. 171), and 'national cluster' (e.g., Porter, 1990, p. 527) are often used interchangeably without differentiation. This lack of clarity leads to misinterpretation and demands further clarification.

### Scale issues

The connection between factors or sources of competitive advantage (Porter, 1990, p. 72) and clustering may lead to falling into the 'scale trap'. Porter discusses the environment in which businesses emerge, evolve, and compete as a facet of competitive advantage. Although it reflects, to some extent, national characteristics, it is not emphasised as national character-istics that are unique to a nation or people. In Porter's context, 'nation' or 'national' refers to a country, but there is an absence of a scale compo-nent. For instance, policy interventions often include the understanding of what national, regional, or local policies have influence on in terms of territory or scope. National economic policy has a country component (specific to a particular country) as well as scale application. However, in industrial clusters the confusion arises, for example, when Porter uses terms like 'national cluster' (Porter, 1990, p. 152) and 'local cluster' (Porter, 1990, p. 171). In the former case, the term refers not to a cluster that cov-ers a whole country but a cluster that is located within a particular coun-try, while in the latter case, it is strongly linked to a limited local scale. Moreover, 'local' often appears to denote national context, particularly when discussing local rivalry in a cluster. It is unclear whether 'local' refers to a country or a cluster in close geographical proximity. Thus, a cluster can be both national and local simultaneously, leading to a mixed geo-graphical scale in discussions. Nonetheless, Porter attempts to integrate the role of local geographical concentration with a country's overall com-petitive advantage, yet this integration lacks proper elaboration.

### Clusters as a concentration vs. a phenomenon

Clusters may be viewed as either a simple concentration of firms, organ-isations, and bodies that are interconnected, or as a phenomenon that enhances the performance of the actors involved. The ambiguity in how Porter describes clusters raises concerns and questions about the concept.

For instance, he notes that 'the process of clustering, and the interchange among industries in the cluster also works best when the industries involved are geographically concentrated' (Porter, 1990, p. 157). However, he does not clarify what the clustering process involves or whether it differs from geographical concentration. The clustering process could be interpreted as an act of gathering or as a specific interchange among a concentrated group of actors. This ambiguity raises further questions about the existence of the phenomenon itself. Furthermore, Porter asserts that 'information is essential to productivity, and relationships that improve its flow will endure and even strengthen after a cluster project ends' (Porter, 1998b, p. 130). However, there is little elaboration on the nature of clusters as projects. If clusters are indeed projects, this implies a different approach, as projects are typically induced, have a limited time-frame, and aim for specific outcomes, which contradicts the essence of Porter's proposed clusters.

### The relationships in clusters

The influence of clusters on competition depends on personal interactions and close contacts. While Porter argues that clusters facilitate the development of such relationships, this does not occur automatically. Proximity can facilitate the development of close relationships, but actors must actively engage in this process (Porter, 1998b). Although Porter (1998b) emphasises the importance of viewing the economy through the lens of clusters, the core benefits of clusters lie in their associations and networking. He claims that the advantages of the cluster stem from personal relationships, open discussions, and trust, with informal contacts and interactions arising from 'living and working in a circumscribed geographic area' (Porter, 1998b, p. 112). The influence of clusters on competition hinges significantly on personal interactions and close contacts, which are critical for fostering innovation and collaboration. It is essential to recognise that such benefits do not materialise automatically, as actors must actively engage in relationship-building and networking.

### Non-obvious facts in the conceptualisation

Porter claims that 'it appears that clusters require a decade or more to develop depth and to gain a real competitive advantage' (Porter, 1998b, p. 119). This observation is evident in numerous examples provided in Porter's works. All the cases have some history of development before being classified as 'clusters', making all of them *ad hoc* clusters. According to Porter (1998b), this is one of the reasons for the failures of government cluster initiatives. Furthermore, Porter states that 'there is no guarantee that a cluster will develop, once the process gets started it is like a chain reaction in which the lines of causality quickly become blurred' (Porter, 1998b, p. 119). This statement not only contradicts the notion that clusters guarantee competitive advantages (Porter, 1990), but also suggests that their development may be left to chance and unclear causal relationships.

*The service application of the concept*

Despite the claims of clusters' importance for both products and services, Porter (1990, 1998b) pays little attention to services. He provides only one example of a tourism cluster (Porter, 1998b, pp. 104–105). However, this example is embedded in wine production and represents a side activity to the wine cluster, which lacks further elaboration. While Porter (1998b) mentions that clusters facilitate the complementarities of products and services, his description is limited to the quality of tourists' experiences, which depend on complementary services around a tourist attraction (Porter, 1998b, p. 110). Overall, the conceptualisation and application of the cluster concept to service providers, in particular to tourism, is underexplored in Porter's work (1990, 1998b).

The cluster literature often refers back to Porter's concept as the starting point of industrial cluster research (e.g., Niu, 2010), although most theoretical frameworks related to the spatial concentration of firms and derived benefits (Asheim, 1996; Becattini, 1990; Landström, 2005; Pyke et al., 1990; Staber, 1998) originate from Marshall (1890/1916), who described some aspects of the concentration and localisation of industries and their benefits. For example, what usually constitutes a source of innovation in a cluster (Desmarchelier & Zhang, 2018; Maté-Sánchez-Val & Harris, 2018; Rahman & Kabir, 2019) relates to the 'hereditary skills' feature of localisation of industries that Marshall (1890/1916) describes. He argues the advantages of hereditary skills by stating that 'the mysteries of the trade become no mysteries; but are as it were in the air, and children learn many of them unconsciously' (Marshall, 1890/1916, p. 271). Marshall does not employ terms such as networking or knowledge spillovers. However, it is common to find Marshall's quote 'in the air' when authors explain the sources of clusters' advantages related to cluster 'atmosphere' (e.g., Bell, 2005; Capone & Lazzeretti, 2018; Fitjar & Rodríguez-Pose, 2017; Porter, 1990).

The industrial cluster concept dates back to Czamanski (Czamanski, 1971, 1974), who used the term and investigated clusters before Porter. Prior to that, such theories as the theory of the location of industries (Weber, 1929), growth poles (orig. 'pôles de croissance') (Perroux, 1950), or industrial complexes (Isard et al., 1959; Isard & Vietorisz, 1955) were introduced. The concepts that emerged later are related and share similar characteristics (Becattini, 2002), including innovative milieus (Aydalot, 1986), industrial districts (Becattini, 1979, 1989), flexible specialisation (Piore & Sabel, 1984; Sabel, 1989), new economic geography (Krugman, 1991), and

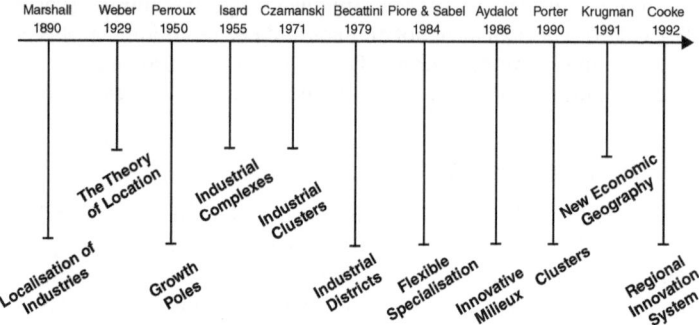

*Figure 2.1* Timeline of the theoretical concepts.

regional innovation systems (Cooke, 1992), among others. Figure 2.1 shows related agglomeration theories introduced at different times. The modern interpretation of clusters has similarities with most theoretical concepts related to business and regional economic development. For example, the concepts of industrial districts (Becattini, 1989) and clusters (Porter, 1990) have much in common. The primary distinction lies in their methodological approaches. While Becattini's (1989, 1990) investigation began with the community, Porter (1990) focused initially on industries, giving less consideration to social relationships. Despite these differences, their practical conclusions are remarkably similar (Landström, 2005).

Due to the obscure conceptualisation of the original cluster concept and its lack of specificity (Martin & Sunley, 2003; Maskell & Malmberg, 2002), there is no clear organisational framework to connect underlying agglomeration-related theories to clusters (Bergman & Feser, 1999/2020). For example, Silicon Valley represents the embodiment of many existing theoretical concepts. Silicon Valley is a network (Ferrary & Granovetter, 2009; Saxenian et al., 2002; Squazzoni, 2009), an industrial district (Cruz & Teixeira, 2010; Gilson, 1999; Markusen, 1996), a growth pole (Rossi, 2009), an example of flexible specialisation (Sabel, 1989), an innovative milieu (Angel, 1991; Pohl & Heiduk, 2002), a regional innovation system (Cooke, 2001), and a cluster (e.g., almost every single article mentions the success of clusters since Porter (1990)). This is a fraction of the research which not only refers to Silicon Valley as an example of the aforementioned concepts but, as noted, also in some cases investigates the valley from these approaches. Silicon Valley has become the classic example of almost any successful concept. However, there is a tendency to ignore, for instance, the defence forces' role in some Silicon Valley firms, with massive budget intakes into these firms, or the fact that the largest employer

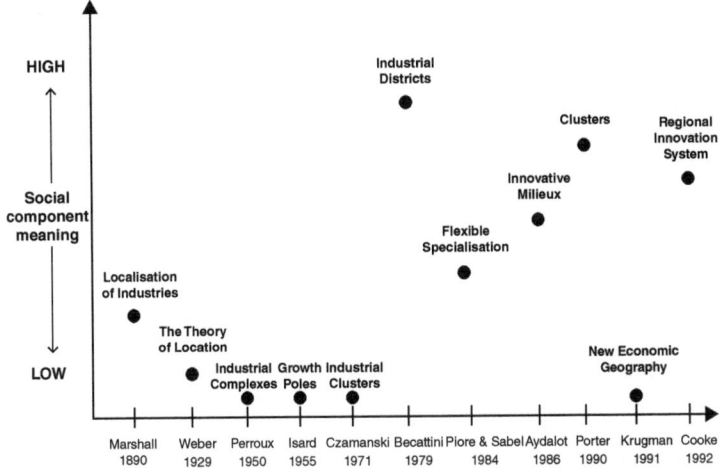

*Figure 2.2* The significance of the social component in theoretical frameworks.

of 16,000 people is a firm that works for the defence sector (Markusen, 2003). Nevertheless, what can be observed is that the social aspect – expressed in the presence of social communities, face-to-face interactions, and tacit knowledge spillovers – has a different degree of importance in different concepts (Figure 2.2). All the theoretical approaches attempt to explain the reasons for co-location or localisation in terms of advantages for firms or industries. While the economic models are self-explained in terms of relying on benefits, they often do not consider the socio-cultural aspects.

The cluster concept has been used as an umbrella concept, covering different agglomeration theories and concepts (Enright, 1996; Vom Hofe & Chen, 2006). It has been applied to numerous industries and sectors, including the textile (Enright, 1996), media (Nachum & Keeble, 2003), creative (Harvey et al., 2012), wine (Giuliani, 2013), wood (Bakarić, 2017), agri-food (Maté-Sánchez-Val & Harris, 2018), aquaculture (Joffre et al., 2019), automotive (Li et al., 2019), and IT (Mittal et al., 2020) industries, among others. Nevertheless, the most common misleading assumptions are that clusters can be identified by formula, that proximity no longer matters in the age of the Internet, that clusters require membership, that clusters are contained within political boundaries, that the public sector can create clusters, and that clusters are a fair and equitable way to grow an economy (Rosenfeld, 2005, pp. 6–10). This situation leads precisely to identifying clusters in an *ad hoc* way (Cortright, 2006; Doeringer & Terkla, 1995) or, for instance, treating clusters as organisations with membership (e.g., Bussmann et al., 2014), or identifying clusters based on managers who showed up at a formal 'cluster' meeting (e.g., Cortright, 2006). The difference between concepts arguably has become so blurred that authors deliberately avoid determining and differentiating between them. Still, the literature on industrial clusters has shown that successful clusters are based on networks comprising firms, people, and institutions (Bathelt, 2002).

### Tourism clusters

The conceptualisation of tourism clusters is similar to the original industrial cluster concept. In a nutshell, a tourism cluster is often defined by the co-location of firms providing tourist services and the existence of relationships between them (among others, Crick et al., 2020; Estevão & Ferreira, 2012; Medeiros, 2021; Miller & Gibson, 2005; Ning, 2021; Novelli et al., 2006). The tourism experience is an aggregate dimension of different products and services consumed during a journey (Novelli et al., 2006; Saxena, 2016), mostly provided by businesses that usually co-locate in a destination (Denicolai et al., 2010; Erkuş-Öztürk, 2009; Tolstad, 2014). It is argued that the overall impression of the destination created requires joint effort and eventually benefits everyone (Nordin, 2003). Collaboration among different tourism businesses is essential to create value for tourists by complementing activities (Mwesiumo & Halpern, 2019), while collaborative marketing assists with the branding of destinations (Fyall & Garrod, 2005).

Since tourism and hospitality businesses are often spatially fixed and tend to co-locate naturally within a destination context (Hall & Williams, 2020), it is constantly proposed that geographical proximity promotes networking and business collaboration (Crick et al., 2020; Fyall et al., 2012; Leick & Gretzinger, 2020). The natural co-location of tourism and hospitality businesses highlights the relevance of the cluster concept in tourism. Being spatially co-located around points of attractions that generate tourist flows, tourism businesses are easily conceptualised as clusters (Martínez-Pérez & Beauchesne, 2018; Wolff et al., 2020). Physical proximity increases casual interactions which develop an awareness of opportunities, knowledge flows, and, consequently, innovations (Gnyawali & Srivastava, 2013). Furthermore, co-location decreases transaction costs and opportunistic behaviour (Martínez-Pérez & Beauchesne, 2018). Spatially fixed businesses are reliant on mutual recognition of shared geographical space, market, and, therefore, interdependence. Since co-located tourism businesses share a level of activity, a physical space, and resources, co-location fosters passive relationships such as competition, collaboration, and joint business activities (McRae-Williams et al., 2007; Taylor et al., 2007). Thus, tourism business co-location is seen to facilitate establishing formal and informal relationships that impact firms' performance.

Tacit knowledge and spillovers play a considerable role in the tourism cluster (Gnyawali & Srivastava, 2013; Jenkins & Tallman, 2010; Raisi et al., 2020). Knowledge spillovers facilitate sustainability, innovativeness, and, eventually, the competitiveness of tourism clusters (Novelli et al., 2006; Shaw & Williams, 2009; Teixeira et al., 2020). Tourism businesses can gain an advantage by accessing knowledge to moderate various aspects of company performance (Martínez-Pérez et al., 2021), identify tourist behaviour trends, design new services (García-Villaverde et al., 2020), develop regional identities and brands (Pongsakornrungsilp et al., 2021), and forge innovations (Leick & Gretzinger, 2020). However, not only do individuals and organisations possess knowledge, it also resides in local networks (Camisón et al., 2017). In other words, close geographical proximity facilitates circulation of knowledge due to collaborative behaviour (Haverniková et al., 2017; Leick & Gretzinger, 2020; Valkokari & Helander, 2007; Zach, 2016; Zach & Hill, 2017).

While the co-location of tourism businesses may be necessary to consider them as a cluster, it does not guarantee economic efficiency or development in and of itself (Huijbens et al., 2014; Perles-Ribes et al., 2017). Just as research shows the positive impacts of forming a tourism cluster in a region (Gardiner & Scott, 2014; Lade, 2010), other tourist destinations successfully develop and compete without being considered as a tourism cluster (Perles-Ribes et al., 2014; Perles-Ribes et al., 2017) or the development of formal cluster programmes does not reflect the actual relationships between tourism businesses in a region (Huijbens et al., 2014). Nevertheless, the cluster concept remains relevant to the development of tourism due to the geographical localisation of tourist activities (Calero & Turner, 2020; Martínez-Pérez et al., 2021; Mwesiumo & Halpern, 2019; Perkins et al., 2022; Teixeira et al., 2020; Zhou et al., 2023).

The intertwinement of economic and social domains is increasingly recognised in tourism cluster literature (Lazzeretti et al., 2019). Despite the claims of the significance of geographical proximity for collaboration (Rodríguez-Victoria et al., 2017), in practice, it is challenging to differentiate clusters and networks (Hall & Williams, 2020). It is argued that clusters and networks differ (Hall, 2004; Nordin, 2003) in so far, at least, as networks designate organisational proximity and clusters represent geographical proximity (Chhetri et al., 2017). Research on tourism clusters in fact explores network structures (e.g., Bittencourt et al., 2019; Chhetri et al., 2017; Jenkins & Tallman, 2010; Raisi et al., 2020) and suggests creating and supporting networks for knowledge exchange to achieve innovativeness and competitiveness (e.g., Binder, 2019; Booyens & Rogerson, 2017; Buffa et al., 2019). Moreover, cluster policies also stress the significance of network relationships within spatial proximity (Rodríguez et al., 2014; Shaw & Williams, 2009). Focusing on networks, the literature emphasises the role of geographical proximity which, in turn, facilitates the development and strength of formal and informal networks.

## Networks

Historically, markets and hierarchies dominated the economy, leading some to argue that networks emerged as an alternative coordination mechanism (Powell, 1991), forming a new economic structure (Cooke & Morgan, 1993; De Man, 2004; Johansson et al., 2012; Molenaar, 2020). Network development has gained significant attention in both academic and government circles, often representing collaborative activities in tourism policy or business. Network theory encompasses various theories that examine the processes connecting network characteristics to desired outcomes (Borgatti & Lopez-Kidwell, 2011). This includes Granovetter's strength of weak ties theory (Granovetter, 1973), social capital theory (Coleman, 1988), and structural holes theory (Burt, 1992).

Networks have long been recognised as a mode of coordinating business activities and as a source of innovation in tourism (Hall, 2004; Hall et al., 1997). The interest in applying and investigating networks in tourism stems from the interconnected and interdependent nature of tourism businesses in delivering experiences to tourists. Tourism is considered as of systemic nature, comprising a set of systems interconnected by tourist flows and external environments (Leiper, 1990). The network framework effectively highlights the systemic nature of tourism, depicting the interdependence of stakeholders and the interconnectedness in the consumption and production processes of tourism products and services.

In general, networks are often interpreted as a set of relationships among tourism-related businesses, organisations, communities, and government, or in a more specific way, such as an industry network, for example, a hospitality network (Gao & Xi, 2018; Möller & Halinen, 2017; Ness et al., 2024). Tourism networks operationalise collaborative behaviour, describing any type of inter-organisational collaborations involving two or more actors. They involve businesses and organisations of all sizes in various combinations that can be domestically or internationally based, and occur at all stages of the value chain. Additionally, destination organisations at different levels, such as national, regional, or local tourism organisations, serve as common organisational structures that facilitate governance of tourism destinations (Volgger & Pechlaner, 2015). Such networks coordinate stakeholders, support tourism development, and market areas both within and outside a country. Furthermore, most initiatives in tourism development begin with the formation of a network organisation intended to represent various tourism businesses and/or a broader set of stakeholders (Martínez-Pérez et al., 2019) in achieving common goals, addressing issues, and coordinating marketing activities for the benefit of businesses, communities, and destinations (Morrison et al., 2004).

Tourism networks provide a broader framework for understanding tourism's functioning and organisation, encompassing both formal and informal types. While formal networks often represent some form of organisation, informal networks are interpreted as social networks closely related to social capital, based on reciprocity, trust, and friendships (Ari et al., 2024; Capone & Lazzeretti, 2018; Kim & Shim, 2018).

Formal networks could take the form of alliances, industry associations, partnerships, and other forms of organisations (van der Zee & Vanneste, 2015). They are characterised by accepting formal terms or signing agreements stating the level of dedication, operational freedom, and independence of individual firms. Tourism and hospitality organisations are often fee-based associations. These forms of networks aim to organise collective representation towards the desired outcome, such as lobbying of interests, information and knowledge exchange, and increasing competitiveness. Also, formal networks are used by policymakers and regional and national tourism organisations to bring tourism and hospitality businesses together to develop attractive regional offerings (Hall, 1999).

Informal networks are often based on the establishment and utilisation of personal connections to engage in business activities without formal terms, obligations, or agreements. Informal relationships prevail in most small tourism business collaborative decisions (Wilke et al., 2019; Zach & Racherla, 2011). Small and medium firms tend to employ informal networks, as personal recognition is prioritised when small firms attempt to establish economic relationships (Teixeira et al., 2019). Frequent informal connections help small firms to gain knowledge and confidence in partners (García-Villaverde et al., 2020). The social networks highlight the application of social capital, which represents a type of resource embedded in social relationships. It is possible to extract benefits out of repeated social interactions due to the values, norms, and relationships formed (Ahn & Ostrom, 2008).

Since collaboration is consistently regarded as significant for tourism businesses and organisations, the network framework is well-suited for understanding collaborative relationships. Destination management and marketing literature sharply focuses on formal and informal networks (Lazzeretti et al., 2019). Destinations themselves have been conceptualised as networks of organisations (Baggio & Cooper, 2010), focusing on collaborative networks and stakeholder collaboration. Collaboration is important for the survival, sustainable performance, and innovation of tourism businesses (Baggio & Valeri, 2022; Binder, 2019; Hall & Williams, 2020). They often represent small- and medium-sized enterprises, so collective action is significant for their survival and development (Martínez-Pérez & Beauchesne, 2018) to offset a lack of resources (Nordin, 2003). From a destination perspective, the collaboration of various tourism businesses is necessary for destination development (Lynch & Morrison, 2006; Perkins et al., 2020).

While various terms are used to conceptualise collaborative relationships among tourism businesses and organisations – including alliances, partnerships, and coalitions (Albrecht, 2013) – the term 'network' is the most commonly used in research and practice regarding tourism development (e.g., among others, Binder, 2019; Grauslund & Hammershøy, 2021; McLeod et al., 2024; Nguyen et al., 2022; Rachmiatie et al., 2024). There is no universally agreed definition for the term 'network', which makes its usage vague (Albrecht, 2013; O'Donnell, 2004). Finding common ground in the utilisation of a precise definition is quite challenging, since networks may span various spatial scales. They may not only take different forms but may also serve different purposes (Tinsley & Lynch, 2001). Also, the objectives of network organisations may include gaining access to broader markets, leveraging specialised and complementary competencies, addressing customer needs, standardising services, and/or securing subsidies or grants (Grabher, 1993; Parra-López & Calero-García, 2009; Ritala & Tidström, 2014). Thus, the definition of network employed will depend on the network's scale, objectives, and core functions. As a result, the term 'network' is commonly used with the broader meaning of networking because the network is often viewed as a structure that supports and facilitates networking (Lynch & Morrison, 2006; Milwood & Roehl, 2018), which, in turn, enables the achievement of goals that would be difficult to accomplish independently (Jesus & Franco, 2016). Thus, any network serves as an access point to a group of businesses and organisations that share similar interests, issues, or resources, offering social networking opportunities to leverage these connections.

The dominant position of networks in the tourism literature is underpinned by the role assigned to networking and social capital as the underlying sources of advantages (Frost & Crockett, 2007). Since business and social relationships get mixed up (Nachum & Keeble, 2003), the social fabric is essential for interactions, knowledge exchange, and mutual identity (Morosini, 2004). Additionally, formal organisations play a role in providing a structure to communicate and share ideas and knowledge among businesses and other relevant actors. Access to information and knowledge through networks is crucial for businesses to adopt ideas and innovate (He & Rayman-Bacchus, 2010; Huang & Wang, 2018; Kajikawa et al., 2012; Liu, 2011). Furthermore, such associations are significant for regional policymaking as they allow a group of businesses to be identified for support programmes. Therefore, networks are considered a vital source of sustainable development in tourism (Lopes et al., 2019; Quaranta et al., 2016) as the process of constant innovation depends on knowledge and information spillovers and their exploitation (Cooper, 2018; Hall & Williams, 2020; Raisi et al., 2020). While the ability to collect, absorb, and exploit knowledge and information differs from firm to firm (Denicolai et al., 2010), quite often the resources for innovation lay outside of a firm (Tinsley & Lynch, 2008). Thus, small- and medium-sized businesses use different networks to access business-related knowledge (Valentina & Passiante, 2009).

Although network organisations are considered significant for knowledge access and diffusion (Foghani et al., 2017; Kim et al., 2023; Yin et al., 2022), geographical proximity remains the essential factor of tacit knowledge diffusion. Moreover, the cluster concept became widely accepted due to the highlighting of network implementation and the significance of local collaboration and community participation (Chin et al., 2017). Tourism clusters are defined as local networks and demonstrate dense network structures (Erkuş-Öztürk, 2009), which are the crucial factor for a successful cluster (Felzensztein et al., 2019; Maghssudipour et al., 2020). Despite the reduction in transportation costs and the increasing speed of information flows, face-to-face interactions remain essential (Breschi et al., 2005; Storper & Venables, 2004). Therefore, the origins of cluster benefits (e.g., among others Enright, 2000; Lechner & Leyronas, 2012) stem from two primary sources that encompass the socio-economic aspects and advantages of clusters: co-location and collaborative networking – the two aspects of clusters which reinforce each other.

The broader societal context is increasingly recognised as integral to economic activities. Networking is a fundamentally interpersonal process, idiosyncratic to the individuals involved. Understanding the embeddedness of economic exchange within social contexts is significant, as certain forms of economic exchange are deeply rooted in social relationships where mutual interests, reputation, and personal connections play significant roles (Powell, 1991). Economic actors do not operate in isolation – they are interconnected through social structures. The social context is continually shaped and reshaped through interactions. Unlike isolated exchanges that focus solely on immediate transactions, these interactions are framed within a broader context that considers past relationships and future outcomes, contributing to the dynamic of business relationships (Grabher, 1993, p. 5). Separating a firm from its societal context and focusing narrowly on economic transactions can overlook how firms operate within a networked environment where social interactions and contextual factors play crucial roles. By contrast, a more integrated perspective recognises that economic activities are deeply embedded in social structures, where networks and relational contexts significantly influence business behaviours. Factors such as reputation, friendship, interdependence, and altruism become essential elements in business relationships (Powell, 1991), influencing collaboration among businesses.

The integration of social aspects into economic exchanges underlines the complexity of economic environments, where both formal and informal relationships shape business behaviours. Incorporating a broader understanding can lead to a comprehensive perspective on how tourism businesses operate and how collaborative behaviours emerge. This perspective aligns with the view that networks are not simply organisational constructs but also social phenomena that shape and are shaped by their environments. Formal networks have the capacity to preserve knowledge, making it more explicit, and facilitating its diffusion among participants. Understanding the economic relevance of networks and the embeddedness of economic activities within social structures highlights the critical role of knowledge exchange and ongoing innovation in tourism, especially for small businesses navigating complex business environments. Thus, the network approach is considered highly relevant to the nature of tourism's functioning (Pforr et al., 2014; Tolstad, 2014).

However, the existence of such networks does not automatically facilitate stakeholders' collaboration (Volgger & Pechlaner, 2015). While the interdependence of tourism businesses plays a vital role in collaborations, organisational, procedural, or relational leadership is also needed (Pechlaner & Volgger, 2012), focusing on specific training and information. Effective collaboration requires more than just network structures. It demands deliberate efforts from governments or business groups to foster engagement and collaboration among participants. However, both formal and informal networks are inherently perceived as collaborative. Based on the proposed network benefits (Morrison et al., 2004), it is assumed that businesses are willing to participate in networks to achieve these benefits (Crick et al., 2020; Fyall et al., 2012). Trust and trustful relationships are often considered the main factor in collaboration and knowledge sharing (Milwood & Roehl, 2018; Strobl & Peters, 2013). The absence of trust complicates networking (Asero et al., 2017; Raisi et al., 2020) and presents an obstacle to knowledge sharing (McComb et al., 2017; Perles-Ribes et al., 2014; Quaranta et al., 2016). However, the absence of trust could simply indicate previous negative experience or an absence of collaboration.

Moreover, since a destination's tourism product comprises numerous businesses, it does not belong to destination organisations. Tourism organisations do not operate or own tourism businesses. Rather, they focus primarily on monitoring and developing tourism. For small tourism businesses, the benefits of network participation might be challenging to quantify. Contributions of businesses to a network organisation can range from membership fees to investments of time and resources. Balancing costs and benefits, particularly for small tourism businesses, is crucial when considering participation in such networks. Thus, networks do not inherently resolve issues of stakeholder alignment, highlighting the importance of proactive measures to facilitate the effective collaboration that is crucial in the context of tourism development. Table 2.1 categorises various tourism network types, detailing functions, descriptions, and examples. This comparison emphasises the importance of local and global networks in tourism, and highlights their spatial distribution with a focus on both geographical and market-specific networks, ranging from community-based networks to online platforms.

*Table 2.1* Tourism network types

| Network type | Description | Function | Examples |
|---|---|---|---|
| Community-based network | Localised networks involving local communities in tourism development and management | Empower local communities, enhance community engagement, promote local culture, and support small tourism businesses | Community tourism boards, village tourism associations, rural tourism networks |
| Regional networks | Networks that connect tourism stakeholders within a particular region, often across multiple towns | Regional marketing, tourism strategy implementation, stakeholder engagement and coordination, and enhancement of regional branding | Regional tourism organisations, destination management organisations, destination marketing organisations |
| National networks | Nationally organised tourism networks, often with government involvement or support | Promote national tourism strategy and enhance international appeal as a tourism destination, often play a significant role in policy coordination or in advocacy and establishing unified standards and strategies for businesses | National tourism organisations (NTOs), national hotel or tour operator networks, national airline associations |
| International and supranational networks | Networks that operate across multiple countries, facilitating cross-border tourism initiatives, standards, and promotions | Facilitate cooperation and coordination in cross-border tourism efforts, promote multiple countries as a destination, increasing cross-border tourism flows, and enhance international regulations, often with a focus on sustainable and responsible tourism, by establishing standards, policies, and joint initiatives | International Air Transport Association (IATA), World Travel & Tourism Council (WTTC), European Travel Commission (ETC) |
| Industry networks | Networks focusing on specific industries within tourism, such as hospitality, accommodation, or tour operators | Provide industry-specific resources, advocacy, and professional development opportunities | Accommodation associations, hospitality associations |

*(Continued)*

*Table 2.1* (Continued)

| Network type | Description | Function | Examples |
|---|---|---|---|
| Specialised networks | Networks formed around specific themes or types of tourism, bringing together businesses and stakeholders with a shared focus, such as adventure tourism, eco-tourism, or wine tourism | Promote niche markets by developing specialised experiences, setting standards, and fostering collaboration among businesses. These networks also help enhance the visibility and appeal of specialised tourism products and services. | Wine tourism network, eco-tourism or sustainable business associations, cultural tourism networks, adventure tourism groups |
| Digital and online networks | Networks based on digital platforms and technologies to connect tourism businesses, tourists, and service providers, enabling interactions, transactions, and information sharing | Facilitate knowledge exchange, reviews, direct marketing, and community building in virtual spaces, particularly the booking and promotion of tourism businesses and services, peer-to-peer recommendations, and customer engagement through social media | Online travel agencies (OTAs) such as Expedia, Booking, Airbnb, or TripAdvisor, influencer communities or social media travel groups such as YouTube travel blogs, Instagram tourism communities, and travel forums |
| Research networks | Networks focused on the exchange of knowledge and research related to tourism, bringing together academics, policymakers, and industry professionals to advance understanding and innovation in tourism | Promote collaboration between researchers and practitioners; support evidence-based tourism policies and strategies; and share insights, trends, and best practices to improve tourism development | Tourism research institutes, academic tourism associations, collaborative research projects, tourism innovation hubs, international academic conferences |

## Ecosystems

As relationships and collaborations become increasingly complex, the concept of business ecosystems has gained momentum, emphasising the dynamics of open systems. The increasingly networked nature of tourism business activities encourages the view of businesses as interconnected entities, leading to collective behaviours resembling those observed in natural ecosystems. The literature highlights the interdependence and collaborative dynamics within business clusters and networks that draw parallels to natural ecosystems (Iansiti & Levien, 2004). The ecosystem approach emphasises the state of these interconnected systems, requiring collaboration among various stakeholders. In this context, resources and processes within a tourism ecosystem are interconnected and do not exist in isolation. Thus, the ecosystem environment is becoming increasingly significant for collaborative behaviour.

The original conceptualisation of business ecosystems possesses similarities to that of industrial cluster: 'Business ecosystems span a variety of industries. The companies within them coevolve capabilities around the innovation and work cooperatively and competitively to support new products, satisfy customer needs, and incorporate ... innovations' (Moore, 1993, p. 15). Marshall (1890/1916, p. xiv) expressed that 'The Mecca of the economists lies in economic biology rather than in economic dynamics. But biological conceptions are more complex than those of mechanics'. Human systems do not exist in a vacuum, and they exhibit indeterministic laws demonstrating changing, unpredictable, non-linear behaviour relative to natural systems (Gunderson & Holling, 2001).

Ecosystems are fundamentally associations of living things and their surroundings, including physical space (Farrell & Twining-Ward, 2004). Geographical proximity is believed to facilitate the creation of such ecosystems, where observation of practices, unconscious learning, trust, norms, and collaborative behaviour shape business behaviours (Cobben et al., 2022; Steinbruch et al., 2022). From this perspective, tourism clusters are argued to form an ecosystem that facilitates collaborative behaviour in various forms (Stål et al., 2023; Vlaisavljevic et al., 2020). Social processes, local communities, and the creation and transfer of knowledge are acknowledged more often as key attributes of successful tourism clusters (Elche et al., 2018; García-Villaverde et al., 2020; Martínez-Pérez et al., 2021). Marshall's understanding that there is something 'in the air' within agglomerations (Marshall, 1890/1916, p. 271) is often used to explain 'atmosphere' in clusters, which leads to competitive advantages (e.g., Bell, 2005; Capone & Lazzeretti, 2018; Fitjar & Rodríguez-Pose, 2017; Porter, 1990). Thus, clusters are recognised as ecosystems within the business context of shared cultural, social, economic, political, and material attributes (Spigel, 2017). The business ecosystem concept represents loosely interconnected networks within clusters, enabling businesses to innovate and adapt quickly in uncertain environments (Auerswald & Dani, 2017; Hui et al., 2022; Lehtonen et al., 2020). In this view, a tourism cluster is an open system that achieves synergy through collaborative behaviour (Vlaisavljevic et al., 2020), facilitated by geographical proximity (Cobben et al., 2022). With increased uncertainty and a turbulent business environment, adopting an ecosystem framework enables flexibility, adaptability, and sustainable growth (Chen et al., 2022).

The business ecosystem concept is used to provide an analogy with ecological systems, characterised by self-organisation, food chain relations, species diversity, and ecological balance (Moore, 1993, 1996, 2006). A business ecosystem was defined as 'an economic community supported by a foundation of interacting organisations and individuals' (Moore, 1996, p. 26). While the term ecosystems and the work of Moore (1996) are often cited in the tourism literature, this does not guarantee that his work has been understood or even read. For instance, as originally proposed, the main feature of the ecosystem concept is the existence of a large company that creates the ecosystem (Moore, 1996), the aspect that is often missing in later discussion of business ecosystems. Moreover, the theorisation of the concept of an ecosystem currently lacks a holistic approach, given that studies tend to focus on the facets of an ecosystem as opposed to its whole (Philipp et al., 2022).

Tourism ecosystems, being complex adaptive systems (Grumadaite, 2020), demonstrate emergent properties (Anggraeni et al., 2007) that lead systems to self-organise (Aarikka-Stenroos & Ritala, 2017). These emergent properties enable the formation of subsystems when specific parameters exceed critical thresholds, enabling the development of new hierarchical levels that simplify overall complexity (Baggio, 2017). Feedback and forward loops are essential for correcting and preventing errors, enabling adaptive change (Farrell & Twining-Ward, 2004). This self-organisation aspect enhances the ecosystem concept's appeal, as it makes it possible to attribute responsibility for events to 'self-organising' features of the system. However, despite the self-organisation attribute, destination ecosystems are approached from the 'how to manage' perspective (Perfetto & Vargas-Sánchez, 2018). Managing a tourism ecosystem is complex, as a firm's impact on the system is limited, and most of the emergent properties depend on the behaviour of numerous actors (Anggraeni et al., 2007). Interactions within the ecosystem often occur at each localised level, where businesses and organisations react to immediate information while remaining largely ignorant of broader system dynamics (Baggio, 2017). However, local interactions make it possible to process information effectively, facilitating rapid learning and the self-organisation of collective behaviour.

To thoroughly understand the dynamics of ecosystems, it is crucial to recognise the importance of cross-scale interactions. A conceptual framework (Figure 2.3) illustrates these interactions among tourism organisations and clusters, as well as their relationship with the broader socio-ecological system. This framework reflects a panarchical approach, emphasising the evolutionary nature of interactions across different scales in both space and time (Holling et al., 2002). Each vertical scale represents a unique adaptive level, while simultaneously reflecting horizontal connections at lower levels, thereby forming an adaptive subsystem within the

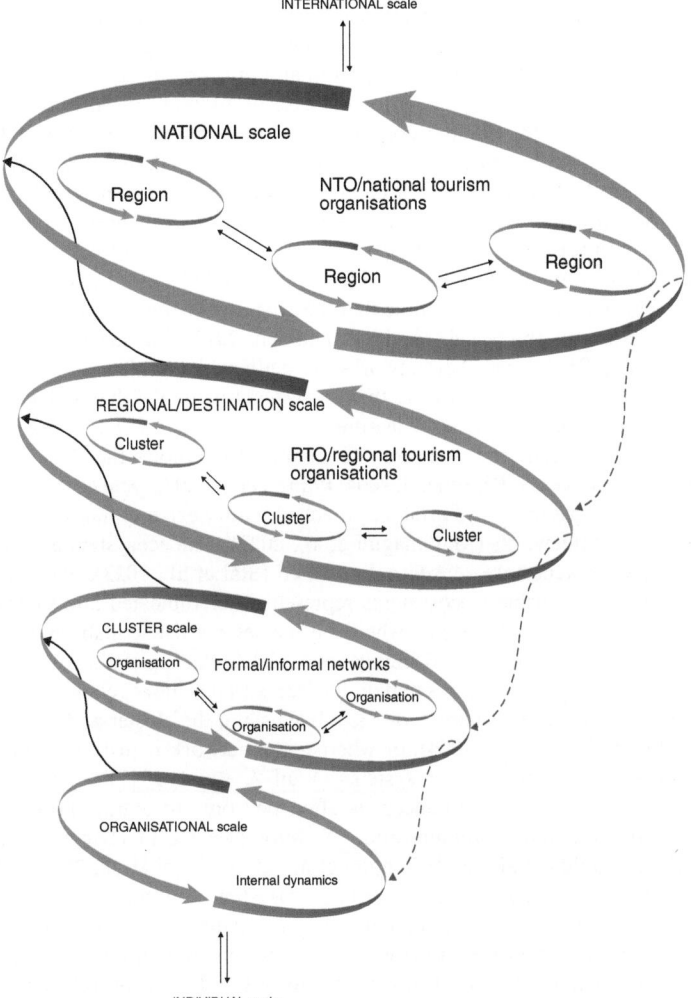

*Figure 2.3* A panarchical structure of interactions within a socio-ecological system.

overarching socio-ecological system. The evolutionary nature of ecosystems reflects the system's ability to evolve and adapt, transitioning up to the next level of self-organisation. The ecosystem approach offers a more holistic perspective that emphasises the interconnectedness of resources and processes among all stakeholders at different levels.

While the ecosystem metaphor is a useful framework, it is potentially problematic for translating, interpreting, and defining human systems without appreciating interactions with, and impacts on, natural systems. This is particularly significant in the case of tourism, given its various impacts. Nevertheless, the concept of ecosystems has been translated without any adjustments to tourism development (Perfetto & Vargas-Sánchez, 2018) and operation involving multiple micro- and small-sized tourism providers. In that sense, it is hard to ignore the emerging associations with tour operators – large companies that create an 'ecosystem' by bundling tourism service providers (e.g., Buhalis & Leung, 2018; Hillebrand, 2022). Critics argue that the ecosystem concept represents the reinvention of the wheel (Möller & Halinen, 2017). For instance, where Powell (1991) argued that networks are neither hierarchies nor markets, Moore (2006) similarly stresses that ecosystems are neither hierarchies nor markets, employing the same features.

In the theorisation, the ecosystem concept resembles the cluster and network concepts (Peltoniemi, 2004; Philipp et al., 2022; Stål et al., 2023). Tourism ecosystems are often approached within existing 'network' and 'cluster' frameworks (Madanaguli et al., 2022). The ecosystem and network approaches are conceptually related (Stål et al., 2023), with some arguing that business ecosystems represent interconnected and interdependent networked systems where businesses coexist, collaborate, and compete with each other (Bachinger et al., 2022; Philipp et al., 2022). The boundaries of the concepts remain unclear, and it is in fact challenging to delimit where the business ecosystem begins and the cluster and network end (e.g., Peltoniemi, 2004), or whether the network is just a structural component of business ecosystems (Wulf & Butel, 2017) because, for example, the ecosystem concept is often tied only to digitalisation and Information and Communication Technologies (ICT) (Baggio, 2020; Giannopoulos et al., 2021; Pohjola et al., 2021; Troisi et al., 2023). The literature tends to utilise the notion of 'ecosystem' as a 'buzzword' (Fuller et al., 2019), while others argue it serves as a panacea for regional development (Scaringella & Radziwon, 2018) which oversimplifies complex issues. For example, the advantages of clusters and networks (Camisón et al., 2017; Novelli et al., 2006; Shaw & Williams, 2009) coincide with the literature on tourism ecosystems (Henche et al., 2020; Morgan et al., 2021; Sedarati et al., 2022; Vargas-Sánchez, 2019). Moreover, some literature directly links business ecosystems and network concepts (e.g., Bachinger et al., 2022). While the notions of 'networks', 'clusters', and 'ecosystems' are used interchangeably in tourism literature (Ogulin et al., 2016), the core mechanism within these concepts is the collaborative behaviour of tourism (Leick & Gretzinger, 2020; Zach, 2016; Zach & Hill, 2017). Table 2.2 outlines the structure of the associational tourism economy, comparing networks (formal and informal), clusters, and ecosystems across different characteristics.

*Table 2.2* Structure of the associational tourism economy, comparing networks, clusters, and ecosystems across key characteristics

| | Networks | | Clusters | Ecosystems |
|---|---|---|---|---|
| | *Informal* | *Formal* | | |
| Governance | No formal governance, self-regulated by participants | Centralised governance; often have local leadership or committee is involved to coordinate activities | Collaborative, often informal but may have local leadership or multiple stakeholders such as local government or private sector leaders, and industry associations | Multi-level, dynamic, and adaptive with multiple layers of governance across public, private, and community stakeholders, often with distributed decision-making |
| Participation | Open, voluntary, based on social ties and shared interests | Membership-based, structured | Geographically bound, arises organically based on shared interest in tourism | Flexible, inclusive of diverse stakeholders with role-based participation at different levels |
| Structure | Loose, flexible, no hierarchy | Often structured hierarchy with clear roles and responsibilities | Semi-structured, based on geographical concentration and interconnectedness with formal and informal networks | Complex, multilayered with interdependent roles among stakeholders |
| Collaboration dynamics | Flexible, voluntary collaborations based on social norms, trust, and mutual benefits | Structured collaborations defined by formal contracts, membership agreements, roles and obligations | Hybrid collaboration, combining informal and formal arrangements within a geographical area | Collaborative dynamics driven by policy agreements, sustainability objectives, and strategic tourism goals |
| Value drivers | Mutual trust, reciprocity, community social capital, local knowledge, and flexibility | Credibility, access to resources, industry standards, and funding | Proximity, shared local focus, resources, and economies of scale and scope | Synergy, sustainability, and resilience across multiple stakeholders |

*(Continued)*

*Table 2.2* (Continued)

| | Networks | | Clusters | Ecosystems |
| --- | --- | --- | --- | --- |
| | *Informal* | *Formal* | | |
| Key benefits | Flexibility, low cost, and rapid adaptation | Stability, trust, access to funding/support, shared resources, structured marketing, and greater influence in policy-making | Enhanced visibility, transaction cost savings, and the creation of unique local experiences | Holistic growth, long-term sustainability, resilience, and shared value creation |
| Common vision | Often implicit, based on shared interests or community goals | Clearly defined, often outlined in mission statements or agreements | Shared local vision focused on collective growth | Broad, long-term vision |
| Knowledge sharing | Fast, informal, often through personal connections and word of mouth | Structured, formal channels such as meetings, reports, and workshops | Continuous exchange, peer-to-peer learning, shared insights, and problem-solving, often through local interactions, events, or joint initiatives | Open and dynamic with broad knowledge shared across diverse stakeholders through collaborative exchange |
| Resilience capacity | High adaptability and community solidarity, allowing quick response to localised challenges but limited long-term stability | Strong stability and institutional support for crisis management and adaptation to market shifts, but less flexible | Collective response to challenges through collaboration, often supported by shared infrastructure, emergency funds, or collective resources | High from diverse support systems, external funding, and policy flexibility for adaptability and sustainable tourism recovery |

## The significance of the geographical dimension

As noted, tourism businesses tend to co-locate and concentrate around tourist attractions or tourist flows. In general, tourism, particularly in rural areas, is spatially fixed and engages local communities (Hall & Williams, 2020). For instance, successful tourism cluster initiatives often involve local communities in planning and organisation, as well as mediating the potential negative effects of tourism (Quaranta et al., 2016). Furthermore, policies stress the importance of spatial proximity to facilitate collaboration for innovation in tourism development (Rodríguez et al., 2014; Weidenfeld et al., 2011). However, the role of spatial proximity remains uncertain in light of the importance of global and local knowledge flows (Quaranta et al., 2016).

Geographical proximity as the physical closeness of businesses and organisations to one another is often considered a critical factor of success, particularly in facilitating networking, knowledge exchange, and collaboration. However, the identification of boundaries is a stumbling block in the network, cluster, and ecosystem concepts. For instance, in attempts to identify different types of clusters, the literature gets to the point where clusters are being applied everywhere (Lazzeretti et al., 2014) and at any scale (e.g., Porter, 1990; Roberts & Enright, 2004; Rosenfeld, 1997). The level of geographical proximity varies, and the exact meaning of proximity remains unclear. A tourism cluster could refer to a broader region (Jackson & Murphy, 2002) or a more localised area (Michael, 2007). In fact, delineating 'cluster boundaries is often a matter of degree, and involves a creative process' (Porter, 1998/2008, p. 104). Similarly, ecosystems and networks can be scale-independent and be established across large distances, as information technologies ease access and communication among actors (Baggio, 2020; Hall & Williams, 2020; Troisi et al., 2023). In general, research tends to promote the generation of networks and knowledge exchange to encourage innovation (e.g., Binder, 2019; Buffa et al., 2019; Martínez-Pérez & Beauchesne, 2018).

While various spatial and temporal dimensions are considered in tourism systems (Hall, 2005), the spatial dimension often remains isolated in tourism research (Farrell & Twining-Ward, 2004). Spatial or temporal factors are often considered secondary, with priority instead being given to issues of trust, reciprocity, and/or social capital (Pohjola et al., 2021; Scaringella & Radziwon, 2018). Nevertheless, it is usually affirmed that geographical co-location is vital for businesses to achieve, at least, agglomeration economies. Moreover, co-location is seen as a facilitator of tacit knowledge spillovers and system feedback (Shaw & Williams, 2009).

Geographical proximity assists in establishing connections among businesses (Crick et al., 2020; Raisi et al., 2020) and enhances the resilience of tourism businesses (Broegaard, 2020; Jang et al., 2021; Rogerson, 2021).

For example, tourism businesses tend to build connections with geographically proximate firms rather than distant counterparts (Raisi et al., 2020). Businesses are likely to have less knowledge about competitors located further away and develop weaker networks with them. Businesses possess greater information about their close rivals and cultivate more developed relationships and networks with them (Crick et al., 2020). Moreover, information and sharing capabilities decay sharply with distance, with substantial effects sometimes occurring within a couple of city blocks in comparison with businesses located ten kilometres apart (Arzaghi & Henderson, 2008; Kerr & Kominers, 2015; Rosenthal & Strange, 2008; Saxenian, 1994). Co-location raises chances for casual and informal interactions, increasing awareness of opportunities, knowledge flow, and innovations within close geographical proximity. For instance, employees may learn about job opportunities within a tourism cluster and more easily transfer from one job to another, facilitating skills and knowledge transfer (Gnyawali & Srivastava, 2013). It is frequently perceived that agglomeration triggers the development of networking (Erkuş-Öztürk, 2009). Consequently, tourism businesses tend to collaborate with nearby counterparts due to the substantial number of unintentional connections, reinforcing the idea of geographical proximity as a driver of collaboration (Kofler et al., 2018). Therefore, geographical proximity is considered essential for establishing formal and informal collaborations for better outcomes.

Co-location is critical for tourism businesses operating within associative concepts. However, simple co-location does not guarantee the success. For example, while it is acknowledged that location plays a significant role in how businesses interact, not all industries and locations interact similarly. Success may differ from region to region even between those with similar attributes (McRae-Williams et al., 2007; Taylor et al., 2007). In some cases, businesses may prefer to exclude government-led organisations from clusters, fearing that the power imbalance could negatively affect potential success (Perkins et al., 2022). Moreover, prioritising collaboration over competition can sometimes be a serious drawback. While competition could pose risks to collaboration and knowledge sharing (Fyall et al., 2012; Raisi et al., 2020), substantial justification should exist to avoid or reduce competition. Although geographical proximity may influence different stakeholders' collaborative behaviour, it can also contribute to competitive behaviours and a reluctance to collaborate.

In general, geographical proximity enhances the observation of competitors and fosters closer relationships with other actors based on both economic interests and the social aspects of belonging to a community or a place. This sense of proximity encourages both formal and informal interactions that foster adaptation, the formation of a shared vision, the establishment of norms, and mutual identification. While co-location may facilitate short-term collaborations, it could also influence longer-term, sustained relationships that evolve over time. Since co-located businesses share a territory, resources, and levels of activity, co-location inherently sustains passive processes in the form of collaboration and competition (Grauslund & Hammershøy, 2021; McRae-Williams et al., 2007; Nguyen et al., 2022; Taylor et al., 2007).

## Conclusion

The concepts of networks, clusters, and ecosystems are central to understanding how tourism businesses, organisations, and communities collaborate and grow. While these terms are often used interchangeably, it is essential to differentiate and understand them to fully comprehend their interconnected functions. Tourism is inherently spatial, as tourists travel through regions, and tourism businesses are spatially fixed. Clusters, representing this geographical concentration, often exist in localised areas, where the proximity of businesses enables more frequent interactions and stronger collaborative opportunities. Networks serve as the underlying mechanisms that facilitate collaboration. Ecosystems encompass a broader combination of factors and environments and their interactions. Therefore, despite the frequent overlap in terminology, it is crucial to recognise that clusters emphasise the geographical concentration of businesses and their interactions, while networks serve as the mechanisms that connect these businesses. Ecosystems, in turn, reflect the state of tourism systems, encompassing both physical and relational dynamics. Whether examining tourism businesses, government interventions, or regional policies, the geographical context must always be acknowledged and integrated into the research and governance of tourism systems. Together, these elements contribute to how tourism systems function, adapt, and thrive.

## References

Aarikka-Stenroos, L., & Ritala, P. (2017). Network management in the era of ecosystems: Systematic review and management framework. *Industrial Marketing Management*, *67*, 23–36. 10.1016/j.indmarman. 2017.08.010

Ahn, T. K., & Ostrom, E. (2008). Social capital and collective action. In D. Castiglione, J. W. Van Deth, & G. Wolleb (Eds.), *The handbook of social capital* (pp. 70–100). Oxford University Press.

Albrecht, J. N. (2013). Networking for sustainable tourism – Towards a research agenda. *Journal of Sustainable Tourism*, *21*(5), 639–657. 10.1080/09669582.2012.721788

Angel, D. P. (1991). High-technology agglomeration and the labor market: The case of Silicon Valley. *Environment and Planning A: Economy and Space*, *23*(10), 1501–1516. 10.1068/a231501

Anggraeni, E., Den Hartigh, E., & Zegveld, M. (2007). *Business ecosystem as a perspective for studying the relations between firms and their business networks*. ECCON 2007 Annual meeting, The Netherlands, Bergen aan Zee.

Araújo, L. (2022). Measuring tourism success: How European national tourism organisations are shifting the paradigm. *Worldwide Hospitality and Tourism Themes, 14*(1), 79–84. 10.1108/WHATT-10-2021-0136

Ari, I. R. D., Prayitno, G., Fikriyah, F., Dinanti, D., Usman, F., Prasetyo, N. E., Nugraha, A. T., & Onishi, M. (2024). Reciprocity and social capital for sustainable rural development. *Societies, 14*(2). 10.3390/soc14020014

Arzaghi, M., & Henderson, J. V. (2008). Networking off Madison avenue. *The Review of Economic Studies, 75*(4), 1011–1038. 10.1111/j.1467-937x.2008.00499.x

Asero, V., Patti, S., & Skonieczny, S. (2017). Cooperative optimization of tourism networks: An application of a game theory model. In P. Vasant, & M. Kalaivanthan (Eds.), *Handbook of research on holistic optimization techniques in the hospitality, tourism, and travel industry* (pp. 348–364). IGI Global.

Asheim, B. T. (1996). Industrial districts as 'learning regions': A condition for prosperity. *European Planning Studies, 4*(4), 379–400. 10.1080/09654319608720354

Auerswald, P. E., & Dani, L. (2017). The adaptive life cycle of entrepreneurial ecosystems: The biotechnology cluster. *Small Business Economics, 49*(1), 97–117. 10.1007/s11187-017-9869-3

Aydalot, P. (1986). *Milieux innovateurs en Europe*. Gremi.

Bachinger, M., Kofler, I., & Pechlaner, H. (2022). Entrepreneurial ecosystems in tourism: An analysis of characteristics from a systems perspective. *European Journal of Tourism Research, 31*. 10.54055/ejtr.v31i.2490

Baggio, J. A. (2017). Simulations and agent-based modelling appendix: Software programs. In R. Baggio & J. Klobas (Eds.), *Quantitative methods in tourism* (pp. 223–244). Channel View Publications. 10.21832/9781845416201-015

Baggio, R. (2020). Digital ecosystems, complexity, and tourism networks. In Z. Xiang, M. Fuchs, U. Gretzel, & W. Höpken (Eds.), *Handbook of e-Tourism* (pp. 1–20). Springer International Publishing. 10.1007/978-3-030-05324-6_91-1

Baggio, R., & Cooper, C. (2010). Knowledge transfer in a tourism destination: The effects of a network structure. *The Service Industries Journal, 30*(10), 1757–1771. 10.1080/02642060903580649

Baggio, R., & Valeri, M. (2022). Network science and sustainable performance of family businesses in tourism. *Journal of Family Business Management, 12*(2). 200–213. 10.1108/JFBM-06-2020-0048

Bakarić, I. R. (2017). The impact of cluster networking on business performance of Croatian wood cluster members. *Croatian Review of Economic, Business and Social Statistics, 3*(2), 39–61. 10.1515/crebss-2017-0008

Bathelt, H. (2002). The re-emergence of a media industry cluster in Leipzig. *European Planning Studies, 10*(5), 583–611. 10.1080/09654310220145341

Becattini, G. (1979). Dal 'settore industriale' al 'distretto industriale'. Alcune considerazioni sull'unità d'indagine dell'economia industrial. *Rivista di Economia e Politica Industriale, 1*, 7–21.

Becattini, G. (1989). Sectors and/or districts: Some remarks on the conceptual foundations of industrial economics. In E. Goodman, J. Bamford, & P. Saynor (Eds.), *Small firms and industrial districts in Italy* (pp. 123–135). Routledge.

Becattini, G. (1990). The Marshallian industrial district as a socio-economic notion. In F. Pyke, G. Becattini, & W. Sengenberger (Eds.), *Industrial districts and inter-firm co-operation in Italy* (pp. 37–51). International Institute for Labour Studies.

Becattini, G. (2002). About the marshallian industrial district and the theory of the contemporary district. A brief critical reconstruction. *Investigaciones Regionales - Journal of Regional Research*, (1), 9–32.

Bell, G. G. (2005). Clusters, networks, and firm innovativeness. *Strategic Management Journal, 26*(3), 287–295. 10.1002/smj.448

Bergman, E. M., & Feser, E. J. (2020). *Industrial and regional clusters: Concepts and comparative applications* (2nd ed.). West Virginia University, Regional Research Institute. (Original work published 1999).

Binder, P. (2019). A network perspective on organizational learning research in tourism and hospitality. *International Journal of Contemporary Hospitality Management, 31*(7), 2602–2625. 10.1108/IJCHM-04-2017-0240

Bittencourt, B., Zen, A., & Prévot, F. (2019). Innovation capability of clusters: Understanding the innovation of geographic business networks. *Review of Business Management, 21*, 647–663. 10.7819/rbgn.v21i4.4016

Booyens, I., & Rogerson, C. M. (2017). Networking and learning for tourism innovation: Evidence from the Western Cape. *Tourism Geographies, 19*(3), 340–361. 10.1080/14616688.2016.1183142

Borgatti, S., & Lopez-Kidwell, V. (2011). Network theory. In J. Scott & P. J. Carrington (Eds.), *The SAGE handbook of social network analysis*. Sage Publications.

Breschi, S., Malerba, F., & Baggio, R. (2005). *Clusters, networks, and innovation*. Oxford University Press.

Broegaard, R. B. (2020). Rural destination development contributions by outdoor tourism actors: A Bornholm case study. *Tourism Geographies*. 10.1080/14616688.2020.1795708

Buffa, F., Beritelli, P., & Martini, U. (2019). Project networks and the reputation network in a community destination: Proof of the missing link. *Journal of Destination Marketing & Management, 11*, 251–259. 10.1016/j.jdmm.2018.05.001

Buhalis, D., & Leung, R. (2018). Smart hospitality – Interconnectivity and interoperability towards an ecosystem. *International Journal of Hospitality Management, 71*, 41–50. 10.1016/j.ijhm.2017.11.011

Burt, R. S. (1992). *Structural holes: The social structure of competition*. Harvard University Press.

Bussmann, U., Panz, R. M., & Schweighofer, S. (2014). *Organisational cultures: Networks, clusters, alliances*. Anchor Academic Publishing.

Calero, C., & Turner, L. W. (2020). Regional economic development and tourism: A literature review to highlight future directions for regional tourism research. *Tourism Economics, 26*(1), 3–26. 10.1177/13548 16619881244

Camisón, C., Forés, B., & Boronat-Navarro, M. (2017). Cluster and firm-specific antecedents of organizational innovation. *Current Issues in Tourism, 20*(6), 617–646. 10.1080/13683500.2016.1177002

Capone, F., & Lazzeretti, L. (2018). The different roles of proximity in multiple informal network relationships: Evidence from the cluster of high technology applied to cultural goods in Tuscany. *Industry and Innovation, 25*(9), 897–917. 10.1080/13662716.2018.1442713

Chandrashekar, D., & Mungila Hillemane, B. S. (2018). Absorptive capacity, cluster linkages, and innovation: An evidence from Bengaluru high-tech manufacturing cluster. *Journal of Manufacturing Technology Management, 29*(1), 121–148. 10.1108/JMTM-05-2017-0087

Chen, M.-K., Wu, S.-W., Huang, Y.-P., & Chang, F.-J. (2022). The key success factors for the operation of SME cluster business ecosystem. *Sustainability, 14*(14), 8236. 10.3390/su14148236

Chhetri, A., Chhetri, P., Arrowsmith, C., & Corcoran, J. (2017). Modelling tourism and hospitality employment clusters: A spatial econometric approach. *Tourism Geographies, 19*(3), 398–424. 10.1080/14616688. 2016.1253765

Chin, W. L., Haddock-Fraser, J., & Hampton, M. P. (2017). Destination competitiveness: Evidence from Bali. *Current Issues in Tourism, 20*(12), 1265–1289. 10.1080/13683500.2015.1111315

Cobben, D., Ooms, W., Roijakkers, N., & Radziwon, A. (2022). Ecosystem types: A systematic review on boundaries and goals. *Journal of Business Research, 142*, 138–164. 10.1016/j.jbusres.2021.12.046

Coleman, J. S. (1988). Social capital in the creation of human capital. *The American Journal of Sociology, 94*. 10.1086/228943

Cooke, P. (1992). Regional innovation systems: Competitive regulation in the new Europe. *Geoforum, 23*(3), 365–382. 10.1016/0016-7185 (92)90048-9

Cooke, P. (2001). Regional innovation systems, clusters, and the knowledge economy. *Industrial and Corporate Change, 10*(4), 945–974. 10.1093/icc/10.4.945

Cooke, P., & Morgan, K. (1993). The network paradigm: New departures in corporate and regional development. *Environment and Planning D: Society and Space, 11*(5), 543–564. 10.1068/d110543

Cooper, C. (2018). Managing tourism knowledge: A review. *Tourism Review, 73*(4), 507–520. 10.1108/TR-06-2017-0104

Cortright, J. (2006). *Making sense of clusters: Regional competitiveness and economic development*. The Brookings Institution.

Crick, J. M., Crick, D., & Tebbett, N. (2020). Competitor orientation and value co-creation in sustaining rural New Zealand wine producers. *Journal of Rural Studies, 73*, 122–134. 10.1016/j.jrurstud.2019.10.019

Cruz, S. C. S., & Teixeira, A. A. C. (2010). The evolution of the cluster literature: Shedding light on the regional studies – Regional science debate. *Regional Studies, 44*(9), 1263–1288. 10.1080/00343400903234670

Czamanski, S. (1971). Some empirical evidence of the strengths of link-ages between groups of related industries in urban-regional complexes. *Papers of the Regional Science Association, 27*(1), 136–150. 10.1007/BF01954603

Czamanski, S. (1974). *Study of clustering of industries.* Institute of Public Affairs, Dalhousie University.

De Man, A. (2004). *The network economy: Strategy, structure and management.* Edward Elgar Publishing.

Denicolai, S., Zucchella, A., & Cioccarelli, G. (2010). Reputation, trust and relational centrality in local networks: An evolutionary geography perspective. In R. Boschma & R. Martin (Eds.), *The handbook of evolutionary economic geography* (pp. 280–297). Edward Elgar Publishing.

Desmarchelier, B., & Zhang, L. (2018). Innovation networks and cluster dynamics. *The Annals of Regional Science, 61*(3), 553–578. 10.1007/s00168-018-0882-5

Doeringer, P. B., & Terkla, D. G. (1995). Business strategy and cross-industry clusters. *Economic Development Quarterly, 9*(3), 225–237. 10.1177/089124249500900304

Elche, D., García-Villaverde, P. M., & Martínez-Pérez, Á. (2018). Inter-organizational relationships with core and peripheral partners in heritage tourism clusters: Divergent effects on innovation. *International Journal of Contemporary Hospitality Management, 30*(6), 2438–2457. 10.1108/IJCHM-11-2016-0611

Enright, M. J. (1996). Regional clusters and economic development: A research agenda. In U. H. Staber, N. V. Schaefer, & B. Sharma (Eds.), *Business networks: Prospects for regional development* (pp. 190–214). De Gruyter. 10.1515/9783110809053.190

Enright, M. J. (2000). Regional clusters and multinational enterprises: Independence, dependence, or interdependence? *International Studies of Management and Organization, 30*(2), 114–138. 10.1080/00208825.2000.11656790

Erkuş-Öztürk, H. (2009). The role of cluster types and firm size in designing the level of network relations: The experience of the Antalya tourism region. *Tourism Management, 30*(4), 589–597. 10.1016/j.tourman.2008.10.008

Estevão, C., & Ferreira, J. J. (2012). Tourism cluster positioning and performance evaluation: The case of Portugal. *Tourism Economics, 18*(4), 711–730. 10.5367/te.2012.0137

Fafurida & Mulyaningsih, T. (2023). A systematic literature review of development rural tourism. *Quality – Access to Success, 24*(194), 35–48. 10.47750/QAS/24.194.05

Farrell, B. H., & Twining-Ward, L. (2004). Reconceptualizing tourism. *Annals of Tourism Research, 31*(2), 274–295. 10.1016/j.annals.2003.12.002

Felzensztein, C., Deans, K. R., & Dana, L. P. (2019). Small firms in regional clusters: Local networks and internationalization in the Southern Hemisphere. *Journal of Small Business Management, 57*(2), 496–516. 10.1111/jsbm.12388

Ferrary, M., & Granovetter, M. (2009). The role of venture capital firms in Silicon Valley's complex innovation network. *Economy and Society*, *38*(2), 326–359. 10.1080/03085140902786827

Fitjar, R. D., & Rodríguez-Pose, A. (2017). Nothing is in the air. *Growth and Change*, *48*(1), 22–39. 10.1111/grow.12161

Foghani, S., Mahadi, B., & Omar, R. (2017). Promoting clusters and networks for small and medium enterprises to economic development in the globalization era. *SAGE Open*, *7*(1). 10.1177/2158244017697152

Frost, M., & Crockett, J. (2007). Communities of practice, clusters or networks? Prospects for collaborative business arrangements in the mining and engineering sector, central western New South Wales. In P. Basu, G. O'Neill, & A. Travaglione (Eds.), *Engagement & change: Exploring management, economic and finance fimplications of a globalising environment* (pp. 249–260). Australian Academic Press.

Fuller, J., Jacobides, M. G., & Reeves, M. (2019). The myths and realities of business ecosystems. *MIT Sloan Management Review*, *60*(3), 1–9.

Fyall, A., & Garrod, B. (2005). *Tourism marketing: A collaborative approach*. Channel View Publications.

Fyall, A., Garrod, B., & Wang, Y. (2012). Destination collaboration: A critical review of theoretical approaches to a multi-dimensional phenomenon. *Journal of Destination Marketing & Management*, *1*(1–2), 10–26. 10.1016/j.jdmm.2012.10.002

Gao, P., & Xi, J. (2018). Network characteristics of tourism destinations: A case from the Yesanpo tourism destination in China. *Journal of Spatial Science*, *63*(2), 245–263. 10.1080/14498596.2018.1465861

García-Villaverde, P. M., Elche, D., & Martínez-Pérez, Á. (2020). Understanding pioneering orientation in tourism clusters: Market dynamism and social capital. *Tourism Management*, *76*, 103966. 10.1016/j.tourman.2019.103966

Gardiner, S., & Scott, N. (2014). Successful tourism clusters: Passion in paradise. *Annals of Tourism Research*, *46*, 171–173. 10.1016/j.annals.2014.01.004

Giannopoulos, A., Piha, L., & Skourtis, G. (2021). Destination branding and co-creation: A service ecosystem perspective. *Journal of Product and Brand Management*, *30*(1), 148–166. 10.1108/JPBM-08-2019-2504

Gilding, M., Brennecke, J., Bunton, V., Lusher, D., Molloy, P. L., & Codoreanu, A. (2020). Network failure: Biotechnology firms, clusters and collaborations far from the world superclusters. *Research Policy*, *49*(2). 10.1016/j.respol.2019.103902

Gilson, R. J. (1999). The legal infrastructure of high technology industrial districts: Silicon Valley, Route 128, and covenants not to compete. *New York University Law Review*, *74*(3), 575–629.

Giuliani, E. (2013). Clusters, networks and firms' product success: An empirical study. *Management Decision*, *51*(6), 1135–1160. 10.1108/MD-01-2012-0010

Gnyawali, D. R., & Srivastava, M. K. (2013). Complementary effects of clusters and networks on firm innovation: A conceptual model. *Journal of Engineering and Technology Management*, *30*(1), 1–20. 10.1016/j.jengtecman.2012.11.001

Gohr, C. F., & Oliveira, I. S. V. D. (2019). Collaboration in cluster-based firms as a source of competitive advantage: Evidence from a footwear cluster. *Production, 29.* 10.1590/0103-6513.20180018

Grabher, G. (1993). *The Embedded firm: On the socioeconomics of industrial networks.* Routledge.

Granovetter, M. (1973). The strength of weak ties. *The American Journal of Sociology, 78*(6), 1360–1380. 10.1086/225469

Grauslund, D., & Hammershøy, A. (2021). Patterns of network coopetition in a merged tourism destination. *Scandinavian Journal of Hospitality and Tourism, 21*(2), 192–211. 10.1080/15022250.2021.1877192

Grumadaite, K. (2020). Sustainable emergence of regional tourism clusters: The approach of complexity theory. *International Conference on Tourism Research.* 10.34190/IRT.20.038

Gunderson, L. H., & Holling, C. S. (2001). *Panarchy: Understanding transformations in human and natural systems.* Island Press.

Hall, C. M. (1999). Rethinking collaboration and partnership: A public policy perspective. *Journal of Sustainable Tourism, 7*(3–4), 274–289. 10.1080/09669589908667340

Hall, C. M. (2004). Small firms and wine and food tourism in New Zealand: Issues of collaboration, clusters and lifestyles. In R. Thomas (Ed.), *Small firms in tourism* (pp. 167–181). Taylor & Francis.

Hall, C. M. (2005). *Tourism: Rethinking the social science of mobility.* Pearson Education.

Hall, C. M., Cambourne, B., Macionis, N., & Johnson, G. (1997). Wine tourism and network development in Australia and New Zealand: Review, establishment and prospects. *International Journal of Wine Marketing, 9*(2), 5–31. 10.1108/eb008668

Hall, C. M., & Williams, A. M. (2020). *Tourism and innovation.* Routledge.

Halme, M. (2001). Learning for sustainable development in tourism networks. *Business Strategy and the Environment, 10*(2), 100–114. 10.1002/bse.278

Harvey, D. C., Hawkins, H., & Thomas, N. J. (2012). Thinking creative clusters beyond the city: People, places and networks. *Geoforum, 43*(3), 529–539. 10.1016/j.geoforum.2011.11.010

Havierniková, K., Lemańska-Majdzik, A., & Mura, L. (2017). Advantages and disadvantages of the participation of SMEs in tourism clusters. *Journal of Environmental Management and Tourism, 8*(6), 1205–1215. 10.14505//jemt.v8.6(22).07

He, Z., & Rayman-Bacchus, L. (2010). Cluster network and innovation under transitional economies: An empirical study of the Shaxi garment cluster. *Chinese Management Studies, 4*(4), 360–384. 10.1108/17506 141011094145

Hemphälä, J., & Magnusson, M. (2012). Networks for innovation – But what networks and what innovation? *Creativity and Innovation Management, 21*(1), 3–16. 10.1111/j.1467-8691.2012.00625.x

Henche, B. G., Salvaj, E., & Cuesta-Valiño, P. (2020). A sustainable management model for cultural creative tourism ecosystems. *Sustainability, 12*(22), 1–21. 10.3390/su12229554

Hillebrand, B. (2022). An ecosystem perspective on tourism: The implications for tourism organizations. *International Journal of Tourism Research, 24*(4), 517–524. 10.1002/jtr.2518

Holling, C. S., Gunderson, L. H., & Peterson, G. D. (2002). Sustainability and panarchies. In L. H. Gundersson & C. S. Holling (Eds.), *Panarchy: Understanding transformations in human and natural systems* (pp. 63–102). Island Press.

Huang, C., & Wang, Y. (2018). Evolution of network relations, enterprise learning, and cluster innovation networks: The case of the Yuyao plastics industry cluster. *Technology Analysis & Strategic Management, 30*(2), 158–171. 10.1080/09537325.2017.1297786

Hui, Y., Jiao, Y., Cui, C., & Ma, K. (2022). Research on innovation ecosystem of dairy industry cluster based on machine learning and improved neural network. *Computational Intelligence and Neuroscience*. 10.1155/2022/4509575

Huijbens, E., Jóhannesson, G., & Jóhannesson, H. (2014). Clusters without content? Icelandic national and regional tourism policy. *Scandinavian Journal of Public Administration, 18*, 63–85.

Iansiti, M., & Levien, R. (2004). *The keystone advantage*. Harvard Business School Press.

Isard, W., Schooler, E. W., & Vietorisz, T. (1959). *Industrial complex analysis and regional development: A case study of refinery-petrochemical-synthetic-fiber complexes and Puerto Rico*. Technology Press of the Massachusetts Institute of Technology.

Isard, W., & Vietorisz, T. (1955). Industrial complex analysis and regional development: With particular reference to Puerto Rico. *Papers in Regional Science, 1*(1), 227–247. 10.1111/j.1435-5597.1955.tb01431.x

Jackson, J., & Murphy, P. (2002). Tourism destinations as clusters: Analytical experiences from the New World. *Tourism and Hospitality Research, 4*(1), 36–52. 10.1177/146735840200400104

Jang, S., Kim, J., & Lee, Y. J. A. (2021). Intertemporal tourism clusters and community resilience. *Professional Geographer, 73*(3), 567–572. 10.1080/00330124.2021.1871768

Jenkins, M., & Tallman, S. (2010). The shifting geography of competitive advantage: Clusters, networks and firms. *Journal of Economic Geography, 10*(4), 599–618. 10.1093/jeg/lbq015

Jesus, C., & Franco, M. (2016). Cooperation networks in tourism: A study of hotels and rural tourism establishments in an inland region of Portugal. *Journal of Hospitality and Tourism Management, 29*, 165–175. 10.1016/j.jhtm.2016.07.005

Joffre, O. M., Poortvliet, P. M., & Klerkx, L. (2019). To cluster or not to cluster farmers? Influences on network interactions, risk perceptions, and adoption of aquaculture practices. *Agricultural Systems, 173*, 151–160. 10.1016/j.agsy.2019.02.011

Johansson, B., Karlsson, C., & Westin, L. (Eds.). (2012). *Patterns of a network economy*. Springer.

Kajikawa, Y., Mori, J., & Sakata, I. (2012). Identifying and bridging networks in regional clusters. *Technological Forecasting and Social Change, 79*(2), 252–262. 10.1016/j.techfore.2011.04.009

Kerr, W. R., & Kominers, S. D. (2015). Agglomerative forces and cluster shapes. *Review of Economics and Statistics*, *97*(4), 877–899. 10.1162/REST_a_00471

Kim, H., Hwang, S.-J., & Yoon, W. (2023). Industry cluster, organizational diversity, and innovation. *International Journal of Innovation Studies*, *7*(3), 187–195. 10.1016/j.ijis.2023.03.002

Kim, N., & Shim, C. (2018). Social capital, knowledge sharing and innovation of small- and medium-sized enterprises in a tourism cluster. *International Journal of Contemporary Hospitality Management*, *30*(6), 2417–2437. 10.1108/ijchm-07-2016-0392

Kofler, I., Marcher, A., Volgger, M., & Pechlaner, H. (2018). The special characteristics of tourism innovation networks: The case of the regional innovation system in South Tyrol. *Journal of Hospitality and Tourism Management*, *37*, 68–75. 10.1016/j.jhtm.2018.09.004

Krugman, P. (1991). Increasing returns and economic geography. *Journal of Political Economy*, *99*(3), 483–499. 10.1086/261763

Lade, C. (2010). Developing tourism clusters and networks: Attitudes to competition along Australia's Murray River. *Tourism Analysis*, *15*, 649–661. 10.3727/108354210X12904412049811

Lai, Y.-L., Hsu, M.-S., Lin, F.-J., Chen, Y.-M., & Lin, Y.-H. (2014). The effects of industry cluster knowledge management on innovation performance. *Journal of Business Research*, *67*(5), 734–739. 10.1016/j.jbusres.2013.11.036

Landström, H. (2005). *Pioneers in entrepreneurship and small business research*. Springer-Verlag. 10.1007/b102095

Lazzeretti, L., Capone, F., Caloffi, A., & Sedita, S. R. (2019). Rethinking clusters. Towards a new research agenda for cluster research. *European Planning Studies*, *27*(10), 1879–1903. 10.1080/09654313.2019.1650899

Lazzeretti, L., Sedita, S. R., & Caloffi, A. (2014). Founders and disseminators of cluster research. *Journal of Economic Geography*, *14*(1), 21–43. 10.1093/jeg/lbs053

Lechner, C., & Leyronas, C. (2012). The competitive advantage of cluster firms: The priority of regional network position over extra-regional networks – a study of a French high-tech cluster. *Entrepreneurship & Regional Development*, *24*(5/6), 457–473. 10.1080/08985626.2011.617785

Lehtonen, M. J., Ainamo, A., & Harviainen, J. (2020). The four faces of creative industries: Visualising the game industry ecosystem in Helsinki and Tokyo. *Industry and Innovation*, *27*(9), 1062–1087. 10.1080/13662716.2019.1676704

Leick, B., & Gretzinger, S. (2020). Business networking in organisationally thin regions: A case study on network brokers, SMEs and knowledge-sharing. *Journal of Small Business and Enterprise Development*, *27*(5), 839–861. 10.1108/JSBED-12-2019-0393

Leiper, N. (1990). Partial industrialization of tourism systems. *Annals of Tourism Research*, *17*(4), 600–605. 10.1016/0160-7383(90)90030-U

Li, S., Han, S., & Shen, T. (2019). How can a firm innovate when embedded in a cluster? – Evidence from the automobile industrial cluster in China. *Sustainability*, *11*(7). 10.3390/su11071837

Lindqvist, G., Ketels, C., & Sölvell, Ö. (2013). *The cluster initiative greenbook 2.0*. Ivory Tower Publishers.

Liu, C.-H. (2011). The effects of innovation alliance on network structure and density of cluster. *Expert Systems with Applications, 38*(1), 299–305. 10.1016/j.eswa.2010.06.064

Lopes, H. D. S., Remoaldo, P., & Ribeiro, V. (2019). Residents' perceptions of tourism activity in a rural North-Eastern Portuguese community: A cluster analysis. *Bulletin of Geography, 46*(46), 119–135. 10.2478/bog-2019-0038

Lynch, P., & Morrison, A. (2006). The role of networks. In E. J. Michael (Ed.), *Micro-clusters and networks: The growth of tourism* (pp. 43–62). Routledge.

Madanaguli, A., Kaur, P., Mazzoleni, A., & Dhir, A. (2022). The innovation ecosystem in rural tourism and hospitality – A systematic review of innovation in rural tourism. *Journal of Knowledge Management, 26*(7), 1732–1762. 10.1108/JKM-01-2021-0050

Maghssudipour, A., Lazzeretti, L., & Capone, F. (2020). The role of multiple ties in knowledge networks: Complementarity in the Montefalco wine cluster. *Industrial Marketing Management, 90*, 667–678. 10.1016/j.indmarman.2020.03.021

Markusen, A. (1996). Sticky places in slippery space: A typology of industrial districts. *Economic Geography, 72*(3), 293–313. 10.2307/144402

Markusen, A. (2003). Fuzzy concepts, scanty evidence, policy distance: The case for rigour and policy relevance in critical regional studies. *Regional Studies, 37*(6–7), 701–717. 10.1080/0034340032000108796

Marshall, A. (1916). *Principles of economics: An introductory volume*. (7th Ed.). Macmillan. (Original work published 1890)

Martin, R., & Sunley, P. (2003). Deconstructing clusters: Chaotic concept or policy panacea? *Journal of Economic Geography, 3*(1), 5–35. 10.1093/jeg/3.1.5

Martínez-Pérez, Á., & Beauchesne, M.-M. (2018). Overcoming the dark side of closed networks in cultural tourism clusters: The importance of diverse networks. *Cornell Hospitality Quarterly, 59*(3), 239–256. 10.1177/1938965517734938

Martínez-Pérez, Á., Elche, D., & García-Villaverde, P. M. (2021). Bridging capital and performance in clustered firms: The heterogeneous effect of knowledge strategy. *Tourism Management, 85*. 10.1016/j.tourman.2020.104264

Martínez-Pérez, Á., Elche, D., García-Villaverde, P. M., & Parra-Requena, G. (2019). Cultural tourism clusters: Social capital, relations with institutions, and radical innovation. *Journal of Travel Research, 58*(5), 793–807. 10.1177/0047287518778147

Maskell, P., & Malmberg, A. (2002). The elusive concept of localization economies: Towards a knowledge-based theory of spatial clustering. *Environment and Planning A, 34*(3), 429–449. 10.1068/a3457

Maté-Sánchez-Val, M., & Harris, R. (2018). The paradox of geographical proximity for innovators: A regional study of the Spanish agri-food sector. *Land Use Policy, 73*, 458–467. 10.1016/j.landusepol.2018.02.024

McComb, E. J., Boyd, S., & Boluk, K. (2017). Stakeholder collaboration: A means to the success of rural tourism destinations? A critical evaluation of the existence of stakeholder collaboration within the Mournes, Northern Ireland. *Tourism and Hospitality Research, 17*(3), 286–297. 10.1177/1467358415583738

McLeod, M., Vaughan, D. R., Edwards, J., & Moital, M. (2024). Knowledge sharing and innovation in open networks of tourism businesses. *International Journal of Contemporary Hospitality Management, 36*(2), 438–456. 10.1108/IJCHM-03-2022-0326

McRae-Williams, P., Lowe, J., & Taylor, P. (2007). The influence of industrial clusters and place on innovation and entrepreneurial behaviour: An empirical analysis of the Australian wine and tourism industries. *The International Journal of Entrepreneurship and Innovation, 8*(3), 165–174. 10.5367/000000007781698590

Medeiros, E. (2021). Development clusters for small places and rural development for territorial cohesion? *Sustainability, 14*(1), 84. 10.3390/su14010084

Michael, E. J. (2007). Micro-clusters in tourism. In E. J. Michael (Ed.), *Micro-clusters and networks: The growth of tourism* (pp. 33–42). Routledge.

Miller, M. M., & Gibson, L. J. (2005). Cluster-based development in the tourism industry: Putting practice into theory. *Applied Research in Economic Development, 2*(2), 47–63.

Milwood, P. A., & Roehl, W. S. (2018). Orchestration of innovation networks in collaborative settings. *International Journal of Contemporary Hospitality Management, 30*(6), 2562–2582. 10.1108/IJCHM-07-2016-0401

Mittal, H., Saurabh, P., Rohit, D., & Mehta, K. (2020). What impedes the success of latemover IT clusters despite economically favorable environments? A case study of an Indian IT cluster. *Technology Innovation Management Review, 10*(1), 54–67. 10.22215/timreview/1321

Molenaar, C. (2020). *The end of competition: The impact of the network economy.* World Scientific Publishing. 10.1142/11608

Möller, K., & Halinen, A. (2017). Managing business and innovation networks – From strategic nets to business fields and ecosystems. *Industrial Marketing Management, 67*, 5–22. 10.1016/j.indmarman.2017.09.018

Moore, J. F. (1993). Predators and prey: A new ecology of competition. *Harvard Business Review, 71*(3), 75–86.

Moore, J. F. (1996). *The death of competition: Leadership and strategy in the age of business ecosystems.* Harper Business.

Moore, J. F. (2006). Business ecosystems and the view from the firm. *Antitrust Bulletin, 51*(1), 31–75. 10.1177/0003603X0605100103

Morgan, M. O., Okon, E. E., Emu, W. H., Olubomi, O. I. E., & Edodi, H. U. (2021). Tourism management: A panacea for sustainability of hospitality industry. *Geojournal of Tourism and Geosites, 37*(3), 783–791. 10.30892/GTG.37307-709

Morosini, P. (2004). Industrial clusters, knowledge integration and performance. *World Development, 32*(2), 305–326. 10.1016/j.worlddev.2002.12.001

Morrison, A., Lynch, P., & Johns, N. (2004). International tourism networks. *International Journal of Contemporary Hospitality Management, 16*(3), 197–202. 10.1108/09596110410531195

Mwesiumo, D., & Halpern, N. (2019). A review of empirical research on interorganizational relations in tourism. *Current Issues in Tourism, 22*(4), 428–455. 10.1080/13683500.2017.1390554

Nachum, L., & Keeble, D. (2003). Neo-Marshallian clusters and global networks: The Linkages of media firms in central London. *Long Range Planning, 36*(5), 459–480. 10.1016/S0024-6301(03)00114-6

Ness, H., Aarstad, J., & Haugland, S. A. (2024). Structural networks and dyadic negotiations in tourism destination ecosystems. *International Journal of Contemporary Hospitality Management, 36*(2), 379–399. 10.1108/IJCHM-03-2022-0309

Nguyen, T. Q. T., Johnson, P., & Young, T. (2022). Networking, coopetition and sustainability of tourism destinations. *Journal of Hospitality and Tourism Management.* 10.1016/j.jhtm.2022.01.003

Nie, P. Y., Wang, C., & Lin, L. K. (2020). Cooperation of firms yielding industrial clusters. *Area.* 10.1111/area.12632

Ning, Y. (2021). Research on the development of Shanghai tourism industry cluster. *9th International Conference on Orange Technology (ICOT), Tainan, Taiwan.* 10.1109/ICOT54518.2021.9680639

Niu, K.-H. (2010). Industrial cluster involvement and organizational adaptation: An empirical study in international industrial clusters. *Competitiveness Review, 20*(5), 395–406. 10.1108/10595421011080779

Nordin, S. (2003). *Tourism clustering & innovation: Paths to economic growth & development.* Etour.

Novelli, M., Schmitz, B., & Spencer, T. (2006). Networks, clusters and innovation in tourism: A UK experience. *Tourism Management, 27*(6), 1141–1152. 10.1016/j.tourman.2005.11.011

Novotná, J., & Novotný, L. (2019). Industrial clusters in a post-socialist country: The case of the wine industry in Slovakia. *Moravian Geographical Reports, 27*(2), 62–78. 10.2478/mgr-2019-0006

O'Donnell, A. (2004). The nature of networking in small firms. *Qualitative Market Research, 7*(3), 206–217. 10.1108/13522750410540218

OECD. (2005). *Business clusters.* 10.1787/9789264007116-en

OECD. (2009). *Clusters, innovation and entrepreneurship.* 10.1787/97892 64044326-en

Ogulin, R., Selen, W., & Houghton, L. (2016). Coordination in a tourism ecosystem: Methods to tackle wicked problems. *Emergence: Complexity and Organization, 18*(1). 10.emerg/10.17357.1f1e70d186bad562e656d3e1 d25c3887

Parra-López, E., & Calero-García, F. (2009). Success factors of tourism networks. In M. Kozak, J. Gnoth, & L. L. A. Andreu (Eds.), *Advances in tourism destination marketing: Managing networks* (pp. 49–61). Routledge. 10.4324/9780203874127-11

Pechlaner, H., & Volgger, M. (2012). How to promote cooperation in the hospitality industry. *International Journal of Contemporary Hospitality Management, 24*(6), 925–945. 10.1108/09596111211247245

Peltoniemi, M. (2004, September). Cluster, value network and business ecosystem: Knowledge and innovation approach. *Paper presented at the conference 'Organisations, innovation and complexity: New perspectives on the knowledge economy'*, University of Manchester.

Pencarelli, T. (2020). The digital revolution in the travel and tourism industry. *Information Technology & Tourism, 22*(3), 455–476. 10.1007/s40558-019-00160-3

Perfetto, M. C., & Vargas-Sánchez, A. (2018). Towards a smart tourism business ecosystem based on industrial heritage: Research perspectives from the mining region of Rio Tinto, Spain. *Journal of Heritage Tourism, 13*(6), 528–549. 10.1080/1743873X.2018.1445258

Perkins, R., Khoo, C., & Arcodia, C. (2022). Stakeholder contribution to tourism collaboration: Exploring stakeholder typologies, networks and actions in the cluster formation process. *Journal of Hospitality and Tourism Management, 52*, 304–315. 10.1016/j.jhtm.2022.07.011

Perkins, R., Khoo-Lattimore, C., & Arcodia, C. (2020). Understanding the contribution of stakeholder collaboration towards regional destination branding: A systematic narrative literature review. *Journal of Hospitality and Tourism Management, 43*, 250–258. 10.1016/j.jhtm.2020.04.008

Perles-Ribes, J., Rodríguez-Sánchez, I., & Ramón Rodríguez, A. (2014). Innovative tourism clusters: Myth or reality? Empirical evidence from Benidorm. 10.2139/ssrn.2424737

Perles-Ribes, J. F., Rodríguez-Sánchez, I., & Ramón-Rodríguez, A. B. (2017). Is a cluster a necessary condition for success? The case of Benidorm. *Current Issues in Tourism, 20*(15), 1575–1603. 10.1080/13683500.2015.1043247

Perroux, F. (1950). Economic Space: Theory and Applications. *The Quarterly Journal of Economics, 64*(1), 89–104. 10.2307/1881960

Perry, M. (2007). From networks to clusters and back again: A decade of unsatisfied policy aspiration in New Zealand. In R. MacGregor & A. Hodgkinson (Eds.), *Small business clustering technologies: Applications in marketing, management, IT and economics* (pp. 160–183). IGI Global.

Pforr, C., Pechlaner, H., Volgger, M., & Thompson, G. (2014). Overcoming the limits to change and adapting to future challenges: Governing the transformation of destination networks in Western Australia. *Journal of Travel Research, 53*(6), 760–777. 10.1177/0047287514538837

Philipp, J., Thees, H., Olbrich, N., & Pechlaner, H. (2022). Towards an ecosystem of hospitality: The dynamic future of destinations. *Sustainability, 14*(2). 10.3390/su14020821

Piore, M. J., & Sabel, C. F. (1984). *The second industrial divide: Possibilities for prosperity*. Basic Books.

Pohjola, T., Gronman, J., & Viljanen, J. (2021, November). Multi-stakeholder engagement in agile service platform co-creation. *44th International convention on information, communication and electronic technology, MIPRO 2021, 1398–1403*. Opatija, Croatia. 10.23919/MIPRO52101.2021.9596665

Pohl, N., & Heiduk, G. (2002). Silicon Valley's innovative milieu: A cultural mix of entrepreneurs/an entrepreneurial mix of cultures? Experiences of european firms. *Erdkunde, 56*(3), 241–252.

Pongsakornrungsilp, S., Pongsakornrungsilp, P., Pusaksrikit, T., Wichasin, P., & Kumar, V. (2021). Co-creating a sustainable regional brand from multiple sub-brands: The andaman tourism cluster of Thailand. *Sustainability, 13*(16). 10.3390/su13169409

Porter, M. E. (1990). *The competitive advantage of nations.* Macmillan.

Porter, M. E. (1998a). *The competitive advantage of nations: With a new introduction* (New ed.). Palgrave. (Original work published 1990)

Porter, M. E. (1998b). *On competition.* Harvard Business School Publishing.

Porter, M. E. (2000). Location, competition, and economic development: Local clusters in a global economy. *Economic Development Quarterly, 14*(1), 15–34. 10.1177/089124240001400105

Porter, M. E. (2008). *On competition.* (Upd. and exp.). Harvard Business School Publishing. (Original work published 1998)

Powell, W. (1991). Neither market nor hierarchy: Network forms of organization. In G. Thomson, J. Frances, R. Levačić, & J. Mitchell (Eds.), *Markets, hierarchies, networks – The coordination of socail life* (pp. 265–276). SAGE Publications.

Pyke, F., Becattini, G., & Sengenberger, W. (1990). *Industrial districts and inter-firm co-operation in Italy.* International Institute for Labour Studies.

Quaranta, G., Citro, E., & Salvia, R. (2016). Economic and social sustainable synergies to promote innovations in rural tourism and local development. *Sustainability, 8*(7), 668. 10.3390/su8070668

Rachão, S., Breda, Z., Fernandes, C., & Joukes, V. (2020). Cocreation of tourism experiences: Are food-related activities being explored? *British Food Journal, 122*(3), 910–928. 10.1108/BFJ-10-2019-0769

Rachmiatie, A., Setiawan, E., Zakiah, K., Martian, F., & Saud, M. (2024). Halal tourism ecosystem: Networks, institutions and implementations in Indonesia. *Journal of Islamic Marketing.* 10.1108/JIMA-09-2023-0286

Rahman, S. M. T., & Kabir, A. (2019). Factors influencing location choice and cluster pattern of manufacturing small and medium enterprises in cities: Evidence from Khulna City of Bangladesh. *Journal of Global Entrepreneurship Research, 9*(1), 1–26. 10.1186/s40497-019-0187-x

Raisi, H., Baggio, R., Barratt-Pugh, L., & Willson, G. (2020). A network perspective of knowledge transfer in tourism. *Annals of Tourism Research, 80*, 102817. 10.1016/j.annals.2019.102817

Ritala, P., & Tidström, A. (2014). Untangling the value-creation and value-appropriation elements of coopetition strategy: A longitudinal analysis on the firm and relational levels. *Scandinavian Journal of Management, 30*(4), 498–515. 10.1016/j.scaman.2014.05.002

Roberts, B. H., & Enright, M. J. (2004). Industry clusters in Australia: Recent trends and prospects. *European Planning Studies, 12*(1), 99–121. 10.1080/0965431031000163 5706

Rodríguez, I., Williams, A. M., & Hall, C. M. (2014). Tourism innovation policy: Implementation and outcomes. *Annals of Tourism Research, 49*, 76–93. 10.1016/j.annals.2014.08.004

Rodríguez-Victoria, O. E., Puig, F., & González-Loureiro, M. (2017). Clustering, innovation and hotel competitiveness: Evidence from the Colombia destination. *International Journal of Contemporary Hospitality Management, 29*(11), 2785–2806. 10.1108/IJCHM-03-2016-0172

Rogerson, J. M. (2021). Tourism business responses to South Africa's Covid-19 pandemic emergency. *Geojournal of Tourism and Geosites, 35*(2), 338–347. 10.30892/GTG.35211-657

Rosenfeld, S. A. (1997). Bringing business clusters into the mainstream of economic development. *European Planning Studies, 5*(1), 3–23. 10.1080/09654319708720381

Rosenfeld, S. A. (2005). Industry clusters: Business choice, policy outcome, or branding strategy? *Journal of New Business Ideas and Trends, 3*(2), 4–13.

Rosenthal, S. S., & Strange, W. C. (2008). The attenuation of human capital spillovers. *Journal of Urban Economics, 64*(2), 373–389. 10.1016/j.jue.2008.02.006

Rossi, U. (2009). Growth poles, growth centers. In Thrift, N. & Kitchin, R. (Eds.) *International encyclopedia human geography* (pp. 651–656). Elsevier.

Sabel, C. F. (1989). Flexible specialisation and the re-emergence of regional economies. In P. Hirst & J. Zeitlin (Eds.), *Reversing industrial decline? Industrial structure and industrial policy in Britain and her competitors* (pp. 17–70). Berg Publishers. 10.1002/9780470712726.ch4

Safonov, A., & Hall, M. (2023). Degrowing rural tourism development: Thinking globally to save the local. In H. Mair (Ed.), *Handbook on tourism and rural community development*. Edward Elgar Publishing.

Saxena, G. (2016). *Marketing rural tourism: Experience and enterprise*. Edward Elgar Publishing. 10.4337/9781784710880

Saxenian, A. (1994). *Regional advantage: Culture and competition in Silicon Valley and Route 128*. Harvard University Press.

Saxenian, A., Motoyama, Y., & Quan, X. (2002). *Local and global networks of immigrant professionals in Silicon Valley*. Public Policy Institute of California.

Scaringella, L., & Radziwon, A. (2018). Innovation, entrepreneurial, knowledge, and business ecosystems: Old wine in new bottles? *Technological Forecasting and Social Change, 136*, 59–87. 10.1016/j.techfore.2017.09.023

Sedarati, P., Serra, F. M. D., & Jakulin, T. J. (2022). Systems approach to model smart tourism ecosystems. *International Journal for Quality Research, 16*(1), 285–306. 10.24874/IJQR16.01-20

Shaw, G., & Williams, A. (2009). Knowledge transfer and management in tourism organisations: An emerging research agenda. *Tourism Management (1982), 30*(3), 325–335. 10.1016/j.tourman.2008.02.023

Sigurðardóttir, I., & Steinthorsson, R. S. (2018). Development of micro-clusters in tourism: A case of equestrian tourism in northwest Iceland. *Scandinavian Journal of Hospitality and Tourism, 18*(3), 261–277. 10.1080/15022250.2018.1497286

Sölvell, Ö. (2009). *Clusters: Balancing evolutionary and constructive forces.* Ivory Tower.

Spigel, B. (2017). The relational organization of entrepreneurial ecosystems. *Entrepreneurship Theory and Practice, 41*(1), 49–72. 10.1111/etap.12167

Squazzoni, F. (2009). Social entrepreneurship and economic development in Silicon Valley: A case study on the joint venture: Silicon Valley network. *Nonprofit and Voluntary Sector Quarterly, 38*(5), 869–883. 10.1177/0899764008326198

Staber, U. (1998). Inter-firm co-operation and competition in industrial districts. *Organization Studies, 19*(4), 701–724. 10.1177/0170840 69801900407

Stål, H. I., Riumkin, I., & Bengtsson, M. (2023). Business models for sustainability and firms' external relationships – A systematic literature review with propositions and research agenda. *Business Strategy and the Environment.* 10.1002/bse.3343

Steinbruch, F. K., Nascimento, L. d. S., & de Menezes, D. C. (2022). The role of trust in innovation ecosystems. *Journal of Business & Industrial Marketing, 37*(1), 195–208. 10.1108/JBIM-08-2020-0395

Storper, M., & Venables, A. J. (2004). Buzz: Face-to-face contact and the urban economy. *Journal of Economic Geography, 4*(4), 351–370. 10.1093/jnlecg/lbh027

Strobl, A., & Peters, M. (2013). Entrepreneurial reputation in destination networks. *Annals of Tourism Research, 40*, 59–82. 10.1016/j.annals.2012.08.005

Swann, G. M. P. (2009). *The economics of innovation: An introduction.* Edward Elgar Publishing.

Taylor, P., McRae-Williams, P., & Lowe, J. (2007). The determinants of cluster activities in the Australian wine and tourism industries. *Tourism Economics, 13*(4), 639–656. 10.5367/000000007782696050

Teixeira, R. M., Andreassi, T., Köseoglu, M. A., & Okumus, F. (2019). How do hospitality entrepreneurs use their social networks to access resources? Evidence from the lifecycle of small hospitality enterprises. *International Journal of Hospitality Management, 79*, 158–167. 10.1016/j.ijhm.2019.01.006

Teixeira, S. J., João, J. M. F., & Correia, R. C. (2020). What do we know about tourism cluster and insular economy: A bibliometric study. *Journal of Spatial and Organizational Dynamics, 8*(2), 107–128.

Tinsley, R., & Lynch, P. (2001). Small tourism business networks and destination development. *International Journal of Hospitality Management, 20*(4), 367–378. 10.1016/S0278-4319(01)00024-X

Tinsley, R., & Lynch, P. A. (2008). Differentiation and tourism destination development: Small business success in a close-knit community. *Tourism and Hospitality Research, 8*(3), 161–177. 10.1057/thr.2008.26

Tolstad, H. K. (2014). Development of rural-tourism experiences through networking: An example from Gudbrandsdalen, Norway. *Norsk Geografisk Tidsskrift*, *68*(2), 111–120. 10.1080/00291951.2014.894561

Troisi, O., Visvizi, A., & Grimaldi, M. (2023). Digitalizing business models in hospitality ecosystems: Toward data-driven innovation. *European Journal of Innovation Management*, *26*(7), 242–277. 10.1108/EJIM-09-2022-0540

UNWTO. (2023). *International tourism highlights, 2023 edition – The impact of COVID-19 on tourism (2020–2022)*. 10.18111/9789284424986

Valentina, N., & Passiante, G. (2009). Impacts of absorptive capacity on value creation. *Anatolia*, *20*(2), 269–287. 10.1080/13032917.2009.10518909

Valkokari, K., & Helander, N. (2007). Knowledge management in different types of strategic SME networks. *Management Research News*, *30*(8), 597–608. 10.1108/01409170710773724

Vargas-Sánchez, A. (2019). The new face of the tourism industry under a circular economy. *Journal of Tourism Futures*, *7*(2), 203–208. 10.1108/JTF-08-2019-0077

Vlaisavljevic, V., Medina, C. C., & Van Looy, B. (2020). The role of policies and the contribution of cluster agency in the development of biotech open innovation ecosystem. *Technological Forecasting and Social Change*, *155*. 10.1016/j.techfore.2020.119987

Volgger, M., & Pechlaner, H. (2015). Governing networks in tourism: What have we achieved, what is still to be done and learned? *Tourism Review*, *70*(4), 298–312. 10.1108/TR-04-2015-0013

Vom Hofe, R., & Chen, K. (2006). Whither or not industrial cluster: Conclusions or confusions? *The Industrial Geographer*, *4*(1), 2–28.

Weber, A. (1929). *Theory of the location of industries*. University of Chicago Press.

Weidenfeld, A., Butler, R., & Williams, A. W. (2011). The role of clustering, cooperation and complementarities in the visitor attraction sector. *Current Issues in Tourism*, *14*(7), 595–629. 10.1080/13683500.2010.517312

Weidenfeld, A., & Hall, C. (2014). Tourism in the development of regional and sectoral innovation systems. In A. A. Lew, C. M. Hall, & A. M. Williams (Eds.), *The Wiley blackwell companion to tourism* (pp. 578–588). 10.1002/9781118474648.ch46

Wilke, E. P., Costa, B. K., Freire, O. B. D. L., & Ferreira, M. P. (2019). Interorganizational cooperation on tourist destination: Building performance in the hotel industry. *Tourism Management (1982)*, *72*, 340–351. 10.1016/j.tourman.2018.12.015

Wolff, G., Wältermann, M., & Rank, O. N. (2020). The embeddedness of social relations in inter-firm competitive structures. *Social Networks*, *62*, 85–98. 10.1016/j.socnet.2020.03.001

Wulf, A., & Butel, L. (2017). Knowledge sharing and collaborative relationships in business ecosystems and networks: A definition and a demarcation. *Industrial Management & Data Systems*, *117*(7), 1407–1425. 10.1108/IMDS-09-2016-0408

Yin, Y., Yan, M., & Zhan, Q. (2022). Crossing the valley of death: Network structure, government subsidies and innovation diffusion of industrial clusters. *Technology in Society, 71*. 10.1016/j.techsoc.2022.102119

Zach, F. (2016). Collaboration for innovation in tourism organizations: Leadership support, innovation formality, and communication. *Journal of Hospitality and Tourism Research, 40*(3), 271–290. 10.1177/109634 8013495694

Zach, F., & Racherla, P. (2011). Assessing the value of collaborations in tourism networks: A case study of Elkhart County, Indiana. *Journal of Travel & Tourism Marketing, 28*(1), 97–110. 10.1080/10548408. 2011.535446

Zach, F. J., & Hill, T. L. (2017). Network, knowledge and relationship impacts on innovation in tourism destinations. *Tourism Management, 62*, 196–207. 10.1016/j.tourman.2017.04.001

van der Zee, E., & Vanneste, D. (2015). Tourism networks unravelled; a review of the literature on networks in tourism management studies. *Tourism Management Perspectives, 15*, 46–56. 10.1016/j.tmp.2015.03.006

Zhou, Q., Qu, S., & Hou, W. (2023). Do tourism clusters contribute to low-carbon destinations? The spillover effect of tourism agglomerations on urban residential $CO_2$ emissions. *Journal of Environmental Management, 330*. 10.1016/j.jenvman.2022.117160

# 3 Collaboration and networking in tourism business

## Introduction

In the rapidly evolving tourism landscape, collaboration and strategic networking are crucial for businesses aiming to achieve sustainable growth and maintain their competitive advantage. Through collaborations, businesses can address challenges that extend beyond individual operations. Whether through local connections or membership in tourism organisations, collaboration fosters knowledge exchange, value creation, and innovation, all of which enhance competitiveness (Hall & Williams, 2020; Lechner & Leyronas, 2012). Aligning with regional or national tourism strategies may particularly be important for tourism businesses in less prominent locations seeking to attract tourists.

Collaboration plays a vital role in destination management, as many aspects of tourism development transcend the capabilities of a single business. On a broader scale, government organisations are more likely to engage with business groups than with individual small enterprises (Perry, 2007). Consequently, small businesses that do not engage in collaboration risk exclusion from decision-making processes, often in favour of larger tourism businesses. For small and micro firms, participation in network organisations provides valuable connections, access to information, and insights into industry dynamics, all of which are essential for their success (Morrison et al., 2004). While collaboration is just one of the strategies that tourism businesses can pursue, it remains essential for the growth and development of businesses (Binder, 2019; Chin et al., 2017; Perkins et al., 2020), contributing to tourism development.

Collaboration for tourism businesses is often viewed through the lens of collaborative marketing, where groups of businesses market their destinations collectively (Fyall & Garrod, 2005). By combining resources, businesses can amplify their appeal, creating a more compelling offering to tourists. Digitalisation has significantly reduced the gap between tourism providers and tourists, enabling small businesses to leverage local specialties in order to attract visitors and stimulate tourism flows (Michael, 2007c),

DOI: 10.4324/9781003293606-3

thus independently capturing a share of the market without relying on large international players. By contrast, large tourism businesses such as major international tour operators or attractions often dominate collaborative processes within a region or a country. These larger stakeholders typically possess the resources to achieve their goals internally and independently. Moreover, they frequently collaborate with regional or national tourism organisations to influence policy and market conditions. Their involvement in these organisations is welcomed due to their role in generating tourist flows, revenue, and employment, which makes them significant to the economy. As a result, large businesses may see little incentive to collaborate with smaller tourism businesses, perceiving limited reciprocal benefits.

However, the challenge lies in involving smaller businesses whose revenue streams may not seem substantial on an individual scale. These businesses typically need to rely on external collaborations to achieve their goals, as their limited internal capabilities necessitate collaborations that offer access to additional knowledge and resources. Among the common challenges faced by small businesses are human resource management, financial management, contract management, and digital marketing. Collaborations with other businesses, industry associations, or tourism organisations can provide valuable support, helping them address these issues. However, many micro and small tourism businesses struggle with collaborations (Ateljevic & Doorne, 2004; Hall, 2004). Unlike larger firms, smaller businesses often lack the necessary resources – such as time, budget, and specialised staff – to actively participate in multiple organisations or build and maintain extensive networks. At the same time, small businesses are often at the heart of rural tourism development initiatives (Pechlaner & Tschurtschenthaler, 2003; Rodríguez et al., 2014). Consequently, collaborations among these smaller businesses are essential for the sustainable development of tourism, particularly in rural markets, as they drive tourists into these areas (Michael, 2007a).

## Tourism and hospitality businesses

While collaboration in tourism primarily emerges in marketing, it is essential to adopt a holistic perspective on collaborative opportunities. Some businesses, including restaurants, vineyards, and local produce factories, may overlook their roles within tourism by not engaging in tourism initiatives, marketing themselves to tourists, or being a part of the regional tourism brand. Literature often distinguishes between tourism and hospitality businesses within various contexts (e.g., Binder, 2019; Lim et al., 2024; Marasco et al., 2018). One viewpoint emphasises that tourism businesses specifically serve tourists, while hospitality, comprising accommodation providers, cafes, and restaurants, focuses on a broader customer base (e.g., Bowie & Buttle, 2011). Based on this assumption, hospitality businesses may argue that they do not belong to tourism business. However, many accommodation providers, often classified as hospitality businesses, primarily cater to travellers seeking temporary lodging. Similarly, some cafes and restaurants serve only individual tourists or organised tourist groups. Furthermore, hospitality associations, which include both accommodation and catering businesses, often struggle to address a variety of needs due to the differing operations, information requirements, and challenges faced by these businesses (Safonov et al., 2023).

From an operational or service perspective, local guests and tourists do not fundamentally differ. For example, a restaurant does not need to treat these groups differently based on the assumption that locals may return while tourists are one-off visitors. A positive experience shared online or with friends can benefit a restaurant or cafe, even if a tourist never returns. It is crucial to recognise that tourists are not limited to those traveling from distant locations, but can also include individuals from neighbouring cities and regions. Therefore, hospitality businesses should not narrowly define their guests. Since both tourists and local residents make choices about where to dine or spend leisure time, hospitality businesses can benefit from catering to both without the need to alter their service procedures. Ignoring tourists equates to overlooking potential growth and development opportunities. Contrarily, relying solely on tourists or significantly shifting focus towards this segment entails risks, as this customer base may not be consistent. Thus, the aim is not to target tourists exclusively but to understand the broader tourism ecosystem in which hospitality businesses operate.

Another perspective views tourism as encompassing destination-level stakeholders, while hospitality comprises more business-specific activities, focusing on provision services to tourists during their trip (e.g., Camilleri, 2018). This distinction highlights the different scales at which the categories of tourism and hospitality are defined. In this view, tourism involves a network of stakeholders – including national tourism organisations, international tour operators, airlines, and government bodies – all working together to develop and promote regional and national economies. Conversely, hospitality is perceived as individual businesses providing services directly to guests, such as accommodations, restaurants, and tourist activities. These businesses are primarily concerned with guest satisfaction and service quality, yet they are integral to the overall tourism experience. Therefore, from this perspective, while tourism represents a broad macro level, focusing on destination-wide businesses, initiatives, and collaborations, hospitality reflects a narrow micro level, concentrating on individual business operations and tourist satisfaction. However, many businesses, such as tourist attractions, museums, and local tour operators offering various activities like rafting, hot air ballooning, and caving, are often categorised as tourism businesses rather than solely hospitality providers.

The third perspective on the distinction between tourism and hospitality emphasises the different focuses of each category (e.g., Slattery, 2002; Yu et al., 2012). Tourism can be viewed through the perspective of the tourist, encompassing the activities, transportation, and accommodation involved in visiting various destinations. By contrast, hospitality represents the business perspective, focusing on the provision of services designed to meet the needs of tourists during their journey. This distinction reflects two viewpoints: (1) the tourist's experience of engaging with the tourism system and (2) the business's role in delivering those experiences and ensuring guest satisfaction. Therefore, when discussing tourism, on the one hand, the focus is on how tourists perceive and interact within tourism systems. On the other hand, hospitality addresses the operational aspects, concentrating on how businesses provide services that contribute to the overall tourism experience. Table 3.1 provides an overview of the distinctions between tourism and hospitality across three key perspectives: customer type, operational level, and focus.

However, the distinction between tourism and hospitality businesses, as outlined in academic literature and by professional organisations, can often complicate a more comprehensive understanding of tourism systems. It is essential to recognise that tourism and hospitality businesses are inherently interdependent, and collectively contribute to the overall appeal of a destination. While differentiation based on customer types, operational levels, or focus may be useful for research purposes, a holistic understanding is necessary for policy and practice.

Understanding the symbiotic relationship between tourism and hospitality is crucial, as is recognising the differentiation that exists between them. While hospitality businesses may not see themselves as part of

*Table 3.1.* Summary of three perspectives on tourism and hospitality

| Distinction | Tourism | Hospitality |
|---|---|---|
| Customer base | Primarily tourists visiting different destinations | A broader customer base, including both tourists and locals |
| Operational level | Destination-wide, involving various stakeholders and activities (e.g., airlines, tour operators, tourism organisations) | Business-specific, focusing on individual businesses (e.g., accommodations, restaurants, cafes) |
| Focus | Tourist's experience and interaction with the tourism system | Business's perspective in providing services and meeting tourist needs |

tourism, their success is often linked to the overall appeal and reputation of a destination (e.g., Buhalis et al., 2023; Kozak & Rimmington, 1998). Tourism can drive more customers to hospitality businesses, while positive experiences at these establishments can enhance the destination's attractiveness. Businesses that believe they do not belong to tourism should recognise their role in tourism. Overlooking these relationships represents a missed opportunity for development and engagement with both tourists and locals. If the goal is to develop effective ecosystems, calls to treat and investigate hospitality separately from tourism should be approached with caution in favour of a more holistic approach.

## Strategic goal setting

Collaboration is a strategic choice shaped by an understanding of desired outcomes and often influenced by the owner or manager's skill set. Determining when to collaborate or compete revolves around business goals and requires consistent networking efforts in both formal and informal settings to maintain visibility, stay informed, and build connections. However, many businesses may often limit their understanding of collaboration to supply contracts (e.g., Song, 2011; Zhang et al., 2009) or participation in organisations for membership fees (e.g., Garrod & Fyall, 2017), missing out on broader collaborative opportunities.

Although tourism is a composite product involving various services, businesses typically pursue their individual interests. Therefore, establishing clear, realistic, and achievable goals aligned with a business's vision and strategic direction is essential (Locke & Latham, 1990, 2013/2017). Without defined objectives, success can become elusive. Quantifying desired outcomes facilitates focused efforts towards achieving them. Without clear goals, meaningful action becomes challenging, and businesses may lose direction and potential growth by focusing solely on guest satisfaction. Thus, when devising business planning strategies regarding competition or collaboration, two critical factors must be considered: (1) the specific goals a business aims to achieve and (2) the means of achieving these goals – whether through internal resources or external collaborations.

### Setting goals

Effective goal-setting is crucial for any successful business, especially in tourism, where trends, seasonality, regional dynamics, and the involvement of various businesses significantly influence development and performance (e.g., Getz & Carlsen, 2000; Son et al., 2021). Central to effective strategic planning is delineating achievable goals that align with operational capabilities. Tourism businesses should categorise goals into short-term, mid-term, and long-term objectives, each requiring distinct timelines (from not exceeding a year to three and five years) and resource allocations for execution (Okumus et al., 2010).

In the short term, tourism businesses often focus on tangible monetary benefits. Goals may include launching promotions, optimising internal procedures, or enhancing service quality. Short-term goals often revolve around immediate needs and operational challenges that require relatively quick resolutions in comparison to other types of goals. In this context, collaboration can take the form of sharing staff, equipment, or promotional campaigns.

Mid-term goals may not yield immediate outcomes but result in delayed benefits, such as enhanced brand reputation or expanded offerings. These goals require businesses to look beyond short-term gains and invest in strategies that strengthen their market position over time. This might involve expanding market reach into new demographics, developing new products, or forming collaborations with other businesses and tourism organisations.

Long-term goals are strategic, aimed at transformative initiatives that ensure sustained success, resilience, and competitive advantage. Examples include establishing a strong brand presence in national and international markets, adopting sustainable practices, or community engagement. From a collaboration perspective, businesses must share visions and commit to collective success, aligning efforts to enhance the overall attractiveness and competitiveness of their destination.

In practice, business goals often encompass a blend of short-, mid-, and long-term objectives, each influencing how businesses position themselves within the collaboration domain. Understanding these temporal dimensions is crucial for prioritising actions and allocating resources effectively. Short-term goals demand immediate attention, requiring rapid resource deployment to capitalise on opportunities or address pressing challenges. By contrast, long-term goals necessitate meticulous planning and sustained commitment, often involving significant resource investments. Breaking down mid- to long-term objectives into smaller, manageable tasks is a practical approach that ensures continuous progress and maintains motivation that can help overcome challenges and adjust strategies as needed.

While the dynamics of collaboration in tourism may vary based on business size and operational scale, aligning operational needs with long-term goals underscores the importance of networking and collaborative efforts in achieving sustainable growth and competitive advantage. Establishing a visible presence within the local business community, fostering relationships with neighbouring businesses, and engaging in collaborative relationships are essential for enhancing guest experiences and operational efficiencies. However, many small tourism businesses struggle to balance immediate needs with long-term strategic goals, as future gains may seem abstract or unattainable when the focus is on day-to-day operations (Getz & Carlsen, 2000). The key question is whether these goals can be achieved independently or if collaboration is necessary. Understanding the timeframe and scope of goals helps determine whether internal resources are sufficient or if external collaborations are needed.

### Sources to achieve goals

#### Internal capacity

Internal capacity refers to the resources and abilities within a business that can be leveraged to meet its goals. If the objectives can be met internally, the focus should be on enhancing internal processes and resources. Improving efficiency may involve increasing staff productivity or optimising workflows (Ashari et al., 2014; Buhalis & Leung, 2018). For larger businesses, this could require collaboration between departments to eliminate redundant tasks and improve communication. However, small tourism businesses often face challenges in this area due to a lack of standardised procedures, making it difficult to establish manageable tasks and maintain consistent service quality. In many instances, external collaboration is essential for tourism businesses to operate effectively.

#### External collaborations

External collaborations are crucial for obtaining opportunities in tourism. Opportunities for external collaboration are abundant, starting with business-to-business relationships and extending to industry organisations – such as restaurant, hospitality, and accommodation associations – which provide valuable information to help businesses improve their operations (Moretti et al., 2024; Volgger & Pechlaner, 2015). Collaboration with nearby businesses can be particularly fruitful, as identifying opportunities within a local geographical area can lead to mutually beneficial collaborations. Relationships in tourism-related organisations can yield valuable knowledge and connections that enhance business offerings. For instance, regional tourism organisations can offer insights into seasonality and strategic planning. Furthermore, while major attractions and natural resources are significant to tourism offerings, they may deteriorate over time due to factors such as anthropogenic pressure or erosion (Scott et al., 2012; Williams & Ponsford, 2009). Businesses that leverage collaboration can adapt to these challenges and create sustainable strategies that enhance their resilience and long-term success.

The strategic advantages of external collaborations extend beyond immediate revenue gains, offering, for instance, enhanced brand visibility, diversified market access, and shared risk mitigation. Businesses can effectively navigate market uncertainties, capitalise on emerging trends, and collectively address challenges such as seasonality, infrastructure development, and regulatory compliance. Joint efforts in competitive tourism markets can strengthen market presence, increase tourist traffic, enhance visitor experiences, and improve overall destination appeal (Fyall et al., 2012; Giuseppe et al., 2022; Volgger & Pechlaner, 2014). Furthermore, achieving sustainable growth in tourism frequently requires collective action, especially among businesses in areas outside major urban centres (Ahn & Ostrom, 2008). Tourism businesses can effectively highlight their unique regional offerings and reach a wider audience of potential tourists. Aligning with regional or national tourism strategies is crucial for businesses aiming to leverage shared resources and compete more effectively in the tourism market.

Engagement with industry associations and tourism organisations is vital for tourism businesses, providing invaluable platforms for networking. These organisations offer access to essential industry insights, market trends, and regulatory updates along with opportunities to connect with peers and stakeholders (Garrod & Fyall, 2017; Zach & Hill, 2017). Participation in industry events, workshops, and conferences not only enhances a business's visibility but also fosters valuable relationships with other businesses. Through these channels, businesses can voice their concerns, influence policy decisions, and collaboratively address industry-wide challenges such as sustainability practices, crisis management, and regulatory compliance. Moreover, local businesses must recognise their integral role within their communities (Tinsley & Lynch, 2008). Collaboration with local organisations is essential for sustainable tourism development, fostering mutual recognition and providing benefits from the rich diversity of cultural heritage. Engaging in community-based tourism initiatives or eco-tourism projects can promote conservation of natural resources and advocate for responsible tourism practices.

### Collaborating with complementary businesses

Collaborations with businesses offering complementary services or products can create synergistic opportunities that enhance the overall tourist experience through cross-promotion, bundled services, and shared resources (Bengtsson & Kock, 2000, 2014). Collaborations among complementary businesses, such as hotels and local tour operators, play a crucial role in enriching visitor experiences. For example, accommodation providers may collaborate with historical sites or museums to offer packages that include guided tours, cultural workshops, or exclusive access to local events. A simple combination of services allows businesses to catch diverse tourists' interests, enhance destination appeal, and encourage longer stays.

Joint marketing initiatives can also increase brand visibility, recognition, and customer engagement. A hotel might partner with a local restaurant or cafe to provide exclusive dining experiences, or promote integrated packages with wellness spas to target niche markets such as food enthusiasts or wellness tourists. Cross-promotions leverage each partner's customer base, social media presence, and marketing channels to reach broader audiences, drive bookings, and foster customer loyalty through value-added experiences.

In terms of operational efficiencies, sharing resources and processes can streamline business operations and optimise service delivery. For instance, a collaboration between accommodation providers and transport services might offer seamless travel solutions, such as accommodation-transport packages for convenient transportation. Such collaborations significantly increase value for tourists, enhancing guest convenience, reduce logistical costs, and help differentiate offerings in competitive markets. However, tourism collaborations are not limited to complementary businesses, as they extend to relationships with competitors (Wang & Krakover, 2008). These partnerships leverage diverse strengths and resources to capitalise on shared market opportunities while navigating competitive challenges collectively.

### Collaborating with competing businesses

Collaboration is a strategic choice and is not always necessary, as competition remains crucial for dynamic development. However, rival tourism businesses, especially those located in close geographical proximity, may also collaborate, for example, to promote their destination or invest in shared projects (Della Corte & Aria, 2016). While collaboration with competing firms is often seen as contradictory to typical business operations, it has great potential for fostering growth and innovation (Crick et al., 2024). Competing businesses can enhance the attractiveness of their location, benefiting from increased tourist flows as they offer opportunities to diversify revenue streams and access new market segments, expanding their market reach. For instance, tourism businesses might collaborate on eco-friendly initiatives or sustainable tourism practices, setting trends that appeal to environmentally conscious travellers (Nguyen et al., 2022). Additionally, competitors may innovate by co-developing new products or services that address evolving tourist preferences. Shared dynamic pricing models, data analytics, and demand forecasting tools are some ways competitors can maintain profitability in highly competitive tourism markets. Through strategic collaboration, competing businesses can create a win-win situation for all stakeholders involved. In that way, competitors can collectively strengthen the competitiveness of the destination, stimulate tourist spending, and support sustainable tourism growth.

Nevertheless, competition, in some cases, may present risks for collaboration (Fyall et al., 2012), and collaboration might sometimes limit competition (Perkins et al., 2020). Interpersonal dynamics play a significant role in business relationships, particularly in small tourism businesses where collaborations are more personal-based. However, despite these interpersonal considerations, economic factors still dominate the collaboration process, with businesses focusing on ensuring that the collaboration is financially viable and that profits are distributed fairly. Equal standards of service quality also play a pivotal role in ensuring successful collaboration, especially when recommending substitute services or packaging joint offerings. However, if personalities clash, these relationships can fall apart, leading to breakdowns in collaborations. Furthermore, conflicts of interest can potentially arise within or between organisations due to interpersonal disagreements. While the financial rationale remains the driving factor for collaborative efforts, particularly among micro and small businesses, personal relationships play a significant role in moderating collaborative behaviour (Safonov et al., 2023).

Although the language and terminology around collaboration may vary, it is generally seen as working together to achieve positive outcomes. When a common issue arises, collaboration is often perceived as a method for businesses to unite against an external problem, such as addressing regulatory issues or market disruptions. However, collaboration is rarely employed to address internal conflicts between businesses. The dominant focus on achieving positive goals rather than resolving conflicts limits the potential for collaboration to act as a conflict resolution process. Even in trustful relationships, collaborations may still fail if conflicting behaviours are not adequately addressed or if businesses refrain from challenging one another regarding questionable behaviour (Kelliher et al., 2018). In close-knit communities, personal relationships are often intertwined with business operations, making conflicts inevitable. While collaboration is typically not considered within the context of conflict resolution, it can in fact be an effective approach to resolving conflicts. Establishing common ground and fostering open communication are crucial for maintaining long-term business success and ensuring that conflicts do not impact the growth of a destination. A collaborative approach to resolving conflicts can prevent enduring issues that could otherwise negatively impact business competitiveness and operations. The reluctance to address conflicts can undermine the success of collaborative initiatives, reinforcing the importance of proactive management within the broader collaboration framework.

While both intense cooperation and intense competition have the potential to cause challenges, the relationships between tourism businesses, particularly those in close geographical proximity, are not merely interconnections or contractual agreements. Rather, these interactions often involve purposeful, collaborative engagements between rivals that go beyond formal agreements. Thus, while competition is essential for fostering dynamic development, tourism businesses often find value in balancing both competition and cooperation.

## Co-opetition

Tourism business success can partially be attributed to the simultaneous combination of competition and cooperation (Bouncken & Kraus, 2013; Felzensztein et al., 2018). The concept of co-opetition, which blends elements of both cooperation and competition, has been a growing area of research. The term 'co-opetition' describes a strategic scenario where competing firms can both win without losing competition (Nalebuff & Brandenburger, 1997). While businesses typically choose to either compete or collaborate, co-opetition presents a framework where both strategies coexist, often leading to mutual benefits. Various interpretations of co-opetition exist (Damayanti et al., 2017), but all emphasise that cooperation and competition can occur simultaneously across different stages, actors, and dimensions of business activities (Kylänen & Rusko, 2011; Rusko, 2024). The balance between two behaviours helps to increase the performance of economic actors. From one point of view, businesses can collaborate to grow the market and then compete to share it (Della Corte & Aria, 2016; Schiavone & Simoni, 2011). From another, co-opetition can emerge among businesses offering substitute services (Gnyawali & Park, 2011).

The timing of competition and cooperation is rarely discussed in the tourism literature (Damayanti et al., 2017). This includes businesses that either compete or cooperate, that have consecutive cooperation and competition relationships, or that cooperate and compete simultaneously. For instance, sequential co-opetition might involve businesses pooling resources to attract tourists to a destination and later competing for their spending (Kylänen & Rusko, 2011; von Friedrichs Grängsjö, 2003; Wang & Krakover, 2008). Simultaneous co-opetition is evident when rivals share resources, such as locations or attractions, or are forced to cooperate due to exogenous shocks such as a natural disaster (Smith & Henderson, 2008).

Small- and medium-sized tourism firms, as well as entrepreneurs, frequently engage in co-opetition to complement one another's services and meet a broad range of tourist needs. An overbooked campsite, for instance, might refer guests to a competing campground, benefitting both parties by ensuring customer satisfaction (Wang & Krakover, 2008). In another example, the antiques trade can adopt co-opetition strategies to make a tourism market more attractive and generate tourism visitation (Michael, 2007c). Complementing products and services, small businesses have the opportunity to develop the local tourism economy by adding more value to the tourists' experience. In that case, competition arises with other places of interests.

Co-opetition is an individual business strategy of collective behaviour designed to create an environment that benefits all stakeholders. This strategy is beneficial because it decreases costs, increases efficiency, and creates a partnership that can benefit everyone involved. Although co-opetition is seen from different perspectives, tourism literature focuses on geographical proximity and co-opetition separately (Bengtsson & Kock, 2014; Fyall et al., 2012; Wang & Krakover, 2008). While networking in geographical proximity facilitates businesses to form joint services and assist in co-opetitive interactions, the co-opetition literature centres around network-based collaborations (Bengtsson & Johansson, 2014; McGrath et al., 2019). Although co-opetition has not significantly been associated with geographical proximity, the spatial concentration of actors, institutions, and organisations in tourism facilitates intentional and unintentional cooperation between rivals (Kylänen & Rusko, 2011; Webb et al., 2021). Despite different motivations, co-location is a strong factor in co-opetition among tourism businesses, allowing for finding common ground, more frequent interaction, and building reciprocal relationships (Grauslund & Hammershøy, 2021). Moreover, co-opetition behaviour has also been linked to ecosystem success (Liu et al., 2023; Moore, 1993; Scaringella & Radziwon, 2018). Thus, co-location is generally recognised as positively contributing to the co-opetition behaviour of tourism businesses.

Co-opetition helps smaller enterprises overcome challenges such as limited resources and knowledge gaps (Devece et al., 2019). Networking plays a crucial role in stimulating co-opetition interactions and shifting behavioural patterns. For instance, van der Zee and Vanneste (2015) emphasise that co-opeting networks form a subset of the broader cluster concept, while Gnyawali and Park (2009) and Rusko (2014) highlight how co-opetition is commonly studied through the lens of networks, including horizontal, vertical, and plural relationships (Bengtsson & Kock, 2014). Informal networks, especially those grounded in socio-economic culture and geographical proximity, can yield positive outcomes (Zach & Racherla, 2011). These networks help balance cooperation and competition, enhancing the performance of both individual actors and the destination itself (Giuseppe et al., 2022; Petrou et al., 2007; Taylor et al., 2007). While public sector support is essential, businesses themselves must be committed to the co-opetition process and have some stimulus, as seen in situations where rivals recommend one another's services when they are overbooked or running low on products, or develop complementary opening hours during off-peak seasons (Kylänen & Mariani, 2012). Rivals are willing to cooperate when they clearly see that the benefits of collaboration exceed the costs of maintaining these competitive relationships. However, the general benefits of co-opetition such as increased tourist numbers, destination appeal, and tourist spending are mainly long-term (Czakon & Czernek, 2016).

Tourism businesses may also demonstrate an unintentional shift in behaviour. While intentional co-opetition is seen as strategic behaviour that involves planning and rational thinking, unintentional co-opetition is based on spontaneous decisions (Kylänen & Rusko, 2011). This similar approach is presented in the form of tacit co-opetition, where co-located competitors cooperate based on trust relationships (Porto-Gomez et al., 2018), with norms and values dictating firms' motivation to cooperate or to compete (von Friedrichs Grängsjö, 2003). These interactions are spontaneous and informal, dependent on a mutual desire to keep the arrangement. The absence of formal contracts helps individual actors to shift from competition to cooperation and vice versa more quickly, responding to fast-changing market conditions (Damayanti et al., 2017). In this sense, community norms and trust form informal institutions and the 'rules of the game'. So, tourism businesses demonstrate co-opetitive behaviour when the interests of actors who share common resources in the competition for tourists intersect. Co-opetition occurs in both explicit and implicit forms, and may extend to inter-group behaviour as well.

However, the benefits of co-opetition can sometimes be offset by challenges. Fear of unequal benefits or costs with competitors may hinder collaboration, and co-opetition can occasionally limit flexibility or lead to opportunistic behaviour (Bouncken et al., 2015; Bouncken & Kraus, 2013; Levy et al., 2017). The interdependence of competitors in co-opetition relationships can negatively affect performance and competitive advantage (Bengtsson & Kock, 2000). Co-opetition might hinder knowledge transfer between destinations, since every region competes for tourist's spending (Werner et al., 2015). Thus, co-opetition is a complex yet promising strategy for tourism businesses, balancing competition and collaboration to enhance destination appeal, attract more tourists, and foster innovation. Whether intentional or spontaneous, co-opetitive behaviour emerges in networks, clusters, and ecosystems, driving growth and sustainability in competitive tourist markets.

## Structured approach to networking

Following the exploration of collaborations with both complementary and competing businesses, maintaining a network becomes essential. Social networks serve as the foundation for collaborative activities and are crucial for achieving business goals, particularly in a competitive environment (Booyens & Rogerson, 2017; Leick & Gretzinger, 2020; Nguyen et al., 2024). Effective personal network management is critical for supporting tourism businesses, so the networking process should therefore be approached systemically.

Social networks represent a structure grounded in both the frequency and quality of interactions (Zhou et al., 2005). This structure forms hierarchical layers, beginning with the 'support clique', a core group of strong relationships with 3–5 people. Beyond this is the 'sympathy group', a layer of 12–15 contacts with whom interaction occurs at least monthly. The next tier, the 'band group', typically comprises 30–50 people who often form identifiable demographic units in time and space (Dunbar, 1993). Members of this layer frequently overlap with the next layer. The 'active network', composed of approximately 150 contacts, includes people with whom one maintains a personal relationship, making an effort to connect at least once every year or two. Thus, the average size of each successive layer is around 5, 15, 50, and 150 contacts (Zhou et al., 2005). Extensions of these layers may reach up to 500 or even 1500 people, but the practical limit for effective social connection is around 150 contacts due to cognitive constraints (Dunbar, 1998). Within this limit, information can flow effectively and remain accessible (Curry & Dunbar, 2013).

A systematic approach to collaboration management focuses on the external environment and relationships beyond immediate business boundaries. It involves using an existing business network and expanding it in alignment with strategic directions. Collaboration requires shared identification, mutual interests, and background knowledge to recognise similarities, including a shared sense of place (Curry & Dunbar, 2013). Thus, since relationship networks must be maintained, it is essential to consider not only what a business aims to gain from other businesses and organisations but also what it can contribute.

Operationalising social networks involves systematically cross-referencing contacts with strategic directions and objectives, assessing feasibility and strategic significance over time, and leveraging connections to achieve desired outcomes. Regular reviews and adjustments of the roles of contacts are necessary to align with evolving business needs and challenges. Revealing such structure and understanding opportunities may enhance business outcomes through structured relationship management despite limitations in actively maintaining a social network (Roberts et al., 2009). This structured approach to managing a network is based on Dunbar's proposition of different layers of social relationships described earlier (Curry & Dunbar, 2013; Dunbar, 1998).

A foundational step in leveraging existing networks is to organise existing contacts based on the closeness and frequency of communication, determining their relevance and potential for support. Contacts may include business peers, close associates with whom a business frequently communicates, individuals or businesses engaged occasionally, or regional stakeholders with whom occasional partnerships can be leveraged for mutual benefit. This network might also consist of potential contacts known but not actively engaged with, but who are relevant to the business environment. This first step helps to establish a baseline of the existing situation, making it possible to leverage current relationships and think strategically about expanding the network. Existing contacts need to be categorised into three distinct categories based on familiarity and frequency of interaction. Figure 3.1 categorises business contacts into close, occasional, and potential circles, highlighting varying interaction frequencies, support levels, and collaboration potential. This structure emphasises the progression from well-established, accessible contacts to untapped potential relationships, illustrating how each contact type can contribute to business objectives.

### *Close contacts*

This circle comprises businesses and individuals with whom regular and substantial interactions occur. These contacts understand the business well, often share similar challenges or goals, and offer direct support or collaboration on various issues. It includes those whom the owner or

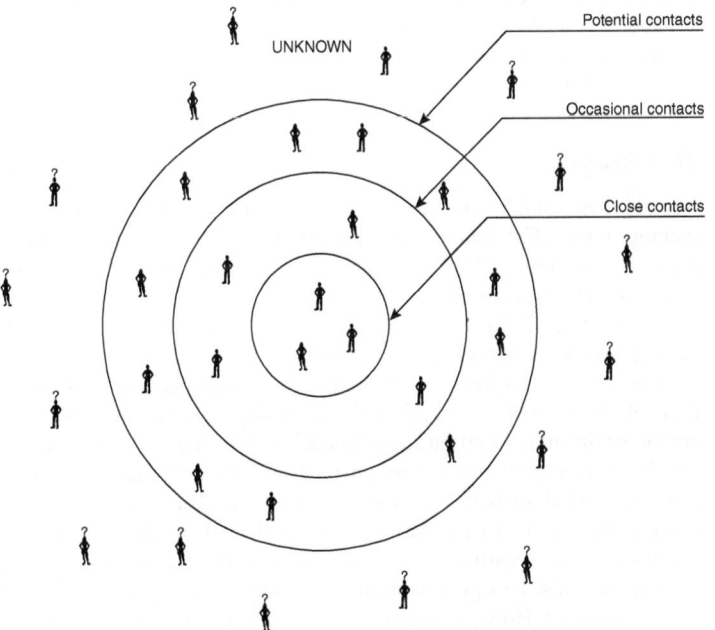

*Figure 3.1* Contact circles for business structured networking.

manager frequently interacts with, both personally and professionally. Due to the frequent communication and mutual support within this group, these contacts are readily accessible, requiring minimal effort to establish and maintain relationships.

### Occasional contacts

This group includes individuals or businesses that are known but with whom interactions occur less frequently. For example, engagement with these contacts may be driven by specific business needs or events. While there is familiarity between both parties, deep relationships may not have developed or may have faded over time. Although these contacts are not as close, they still offer potential for collaboration that can be leveraged for mutual benefit. Establishing communication with this circle requires a bit more effort to pursue business goals.

### Potential contacts

This group consists of individuals or organisations that may be known and relevant to the business or its location, but with whom there has been little to no active engagement. These contacts represent opportunities for collaborations that could support the achievement of business goals. Nurturing these relationships is essential to transition them into more active collaborative partners.

### The unknown

Beyond these established categories lies the realm of the unknown, which includes unfamiliar contacts and opportunities in new markets, associations, or businesses. These organisations represent untapped potential that could significantly contribute to business growth if effectively discovered and engaged. Therefore, uncovering and exploring these opportunities is essential for achieving business objectives.

The structured approach to managing personal networks can be further refined to support specific business goals, such as cross-referencing contacts with individual business objectives. This method helps evaluate whether current contacts can support a given goal or if additional strategies are needed to better leverage the network. Transitioning contacts from unfamiliarity to familiarity, moving them from the periphery to closer circles, is essential for nurturing meaningful relationships focused on specific goals. This process strengthens relationships that contribute to business success. However, networking abilities have limitations regarding the number of connections people can actively engage with (Roberts et al.,

2009). For example, networking requires proactive communication, staying informed on activities, and sustained engagement, which in turn demands emotional involvement that diminishes across the larger network layers (Dunbar, 1998; Hill & Dunbar, 2003). Thus, understanding a business network and its relevance to business objectives is vital for allocating resources effectively.

### *Revisiting and adapting the framework*

Networks are dynamic and require periodic review and adjustment to stay aligned with evolving business goals and external conditions. Regular reassessment ensures the network remains relevant and capable of supporting current objectives. Collaboration with other businesses or related organisations can be complex, often requiring reciprocity to ensure mutual benefit. Optimising contact engagements according to objectives enhances resource allocation and maximizes the impact of collaborations. This strategic networking approach should be integral to any business scenario, recognising that tourism businesses do not operate in isolation. Businesses must continually reassess their relationships and strategies. This ongoing adaptation is crucial for sustaining growth and competitiveness in tourism. A structured approach to managing and leveraging personal relationships is essential to achieve business objectives through a comprehensive understanding of the business network.

### Non-collaboration of tourism businesses

Networking is often highlighted in the literature as a core driver of competitive advantage in tourism development frameworks (Hoholm, 2015; Tunisini & Marchiori, 2020; Volgger & Pechlaner, 2015). Both academic research and policy typically assume that businesses are willing to participate in collaborations, motivated by the potential benefits. However, successful collaborations are relatively rare (Pforr, 2004), and the risks associated with networking can discourage businesses from willingly engaging in collaborations (Pechlaner & Volgger, 2012). It is common to assume that expanding markets may incentivise businesses to share knowledge and join networks (e.g., Fyall et al., 2012). While businesses may seek valuable contacts or access to politicians and lobbyists, they often remain hesitant when it comes to sharing proprietary information. Small businesses in rural areas in particular face additional obstacles when attempting to engage in collaborations (Leick & Gretzinger, 2020). While networking is promoted as essential for its potential benefits, noncollaborative behaviour remains prevalent among businesses and organisations in tourism.

Actors may avoid collaborative interactions to protect knowledge, minimise resource wastage, cut their costs, or keep opportunities open (Hoholm, 2015). Geographical factors can also influence collaboration, as unique local characteristics are often challenging to replicate in distant collaborations (Gilding et al., 2020). Even in close geographical settings, such as science parks, businesses may refrain from collaborating if they lack things in common (Ruokolainen & Igel, 2022). Bureaucratic barriers and conflicts of interest further inhibit collaboration among universities, government entities, and businesses (Chin et al., 2017). Additionally, rural businesses may avoid formal associations due to concerns over potential urban dominance in regional networks (Pilving et al., 2022). Membership fees can also raise expectations for returns, which may in turn overshadow the core objectives of formal networks (Phillipson et al., 2006). Local communities might similarly avoid participation in network organisations, perceiving them as power structures that negatively impact decision-making and tourism policy implementation at the local level (Saufi et al., 2014). External interventions that overlook local context can disrupt established norms and networks, potentially damaging the social capital structures within a community (Erkuş-Öztürk, 2011; Phillipson et al., 2006). Social capital, a resource embedded in social network relationships, plays a vital role in supporting collaboration through the trust and shared norms established by repeated interactions (Lin, 2008). Broadly, social capital encompasses values, norms, and relationships developed over time to help address social dilemmas (Ahn & Ostrom, 2008). In tourism, social capital reflects the social structures that facilitate collaboration (Kim & Shim, 2018).

A lack of trust is commonly cited in tourism literature as a primary obstacle to collaboration (Leick & Gretzinger, 2020; McComb et al., 2017; Perles-Ribes et al., 2014; Quaranta et al., 2016). Trust is considered essential for fostering collaboration and knowledge sharing, and its absence can significantly hinder these processes (e.g., Raisi et al., 2020). Successful collaborations, especially in clusters, networks, and ecosystems approaches, rely heavily on trust, particularly among businesses that co-exist and evolve in close geographical proximity (Kelliher et al., 2018; McTiernan et al., 2021; Scaringella & Radziwon, 2018). However, basic transactions do not require high-trust relationships (Cooke & Morgan, 1998). When the cost (and also the value) of such relationships is low, they could be easily replaced or disappear if they do not impact operations at all. By contrast, high trust is necessary for strategic or high-cost relationships. While trust is often cited as significant in relationships to deal with uncertainties and reduce transaction costs, distrust was also found to be necessary for performance in co-opetition relationships (Raza-Ullah & Kostis, 2020). Beyond trust, other factors can influence non-collaborative behaviour. Large businesses, for instance, may lack interest in collaborating with smaller counterparts, and reluctance to engage can also stem from an absence of collective punitive actions towards self-serving or opportunistic behaviour (Perry, 2001). Moreover, even in close geographical proximity, businesses may vary in their economic, cultural, and organisational proximities, which can significantly impact their collaborative behaviour (Sørensen, 2007). While limited trust is often noted, it alone does not provide a complete explanation for non-collaborative behaviour.

Trust is a complex phenomenon (Steinbruch et al., 2022) and cannot be demanded from participants (Scaringella & Radziwon, 2018). Thus, network, cluster, or ecosystem frameworks aimed at facilitating interactions could bump into the inability to stimulate trust, eventually leading to firms quitting the programmes when funding finishes (Ingley, 2008), or if not quitting, then weakening collaboration linkages (Rodríguez et al., 2014). Tourism policies tend to overthink their significance for businesses and become overly ambitious in delivering development programmes without consideration of participants' needs (Ingley, 2008; Nishimura & Okamuro, 2011; Perry, 2007; Rodríguez et al., 2014). As with many collaborative projects, networks often fail to reach their potential or dissolve entirely. In some cases, the role of institutional stakeholders is less important to businesses' internal capabilities and mutual recognition of partners (Tunisini & Marchiori, 2020). A lack of belief in the network's value or insufficient cohesion among participants can quickly lead to network failure. Generating trust requires time, willingness to believe in future gains, and deep grounds for understanding mutual interdependence. Businesses frequently overlook the potential benefits, remaining reluctant to engage fully in the opportunities a network could offer (Gilding et al., 2020; Hoholm, 2015). Collaboration is more effective when it is based on some shared 'history' of interactions formed by common territorial identity (Rodríguez et al., 2014). Some argue that opportunism is a key factor behind network failures, as individual participants may prioritise their own gains over collective success (Schrank & Whitford, 2011). However, when an association is a fee-based entity that often promises benefits in exchange (Phillipson et al., 2006), the ways in which opportunism is operationalised within such networks are rarely acknowledged. Thus, rather than attempting to force interactions through a formal association and monetary incentives, understanding local context to facilitate the sense of belonging to the higher strategic purpose, norms, and common culture may potentially achieve better outcomes.

Tourism businesses generally do not realise that knowledge and information are a source of competitiveness and are therefore unwilling to invest their resources in initiatives of this kind (Raisi et al., 2020). Also, collaborative relationships are not straightforward and, in the way they are generally suggested, require a certain degree of altruism. However, close relationships often come with drawbacks such as loss of control, reinterpretation of relationships and behaviour, the need for additional resources, and prioritising one or other business or association. These drawbacks are bound to arise sooner or later and can only be mitigated but not avoided (Håkansson & Snehota, 1995). Furthermore, collaboration is complicated by the complexities of relationships, issues with identifying stakeholders' different interests and goals, and issues with involving particular stakeholders in collaboration (Perkins et al., 2020). The current state of tourism literature and policy fails to recognise different perspectives, especially the existence of non-collaborative behaviour among businesses with regard to their participation in associations.

Attention has been paid mainly to network structures and dynamics, assuming the positive impact of networking (Tunisini & Marchiori, 2020). However, assuming that interactions always have positive connotations is naive and dangerous for businesses and policymakers (Hoholm, 2015). Nevertheless, if interdependence is considered, relationships are not so unequivocal. Furthermore, defining interdependence in tourism is challenging due to the interpretation of tourism as being an industry. A tourism business is often defined as any firm that supplies goods and services to tourists. However, such an approach simplifies the reality, which could lead to unsuccessful policies and collaboration initiatives. This approach is often based on the demand side of activities and expenditure of tourists that supposedly represent a tourism industry (Leiper et al., 2008). However, if the emphasis is placed on the supply side, tourism may be considered 'partially industrialised' (Leiper, 1990). Therefore, businesses that are not considered tourism businesses may not collaborate with other co-located businesses or be considered important by tourism organisations.

While informal and formal networks, clusters, and ecosystems share certain similarities, they differ in the level of formality and the stability of their relationships. Formal networks are characterised by relatively stable, structured relationships, often based on formal membership agreements and consequent fees. Clusters encompass both formal and informal relationships, with varying degrees of stability depending on the mix of these interactions. Informal networks, meanwhile, are primarily based on non-economic exchanges and social interactions, leading to less stability in economic terms. However, they play a crucial role in establishing recognition and association among members. An ecosystem represents a state within this broader system, integrating informal networks, formal networks, and clusters into a dynamic, interdependent environment. While individual networks or clusters can function as ecosystems on their own, the ecosystem as a whole reflects the broader, interconnected state of these entities. Figure 3.2 illustrates how these categories of the associational economy intersect, highlighting distinctions in relationship types and their durability.

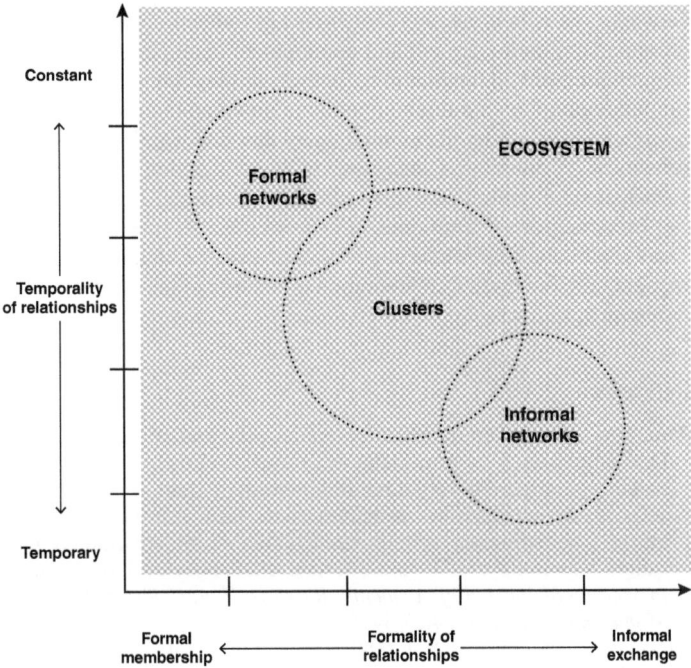

*Figure 3.2* Relationships in the associational tourism economy based on formality and temporality of relationships.

## Conclusion

The successful navigation of tourism complexities hinges on effective goal-setting, internal capacity evaluation, and effective collaborations. Collaborating with complementary businesses can lead to mutual growth, while even competing businesses can find common ground on specific issues. Collaborative relationships within a geographical area extend beyond simple market exchanges, creating an ecosystem of interconnected and interdependent relationships. While collaboration can help resolve local conflicts and promote a thriving business environment, it is essential to maintain a balance between collaboration and competition. Businesses must set clear goals and understand how to achieve them. Assessing internal capacity helps determine whether goals can be met with existing resources or if external collaborations are necessary. For small businesses, external collaborations are often crucial for growth and success.

A structured approach to networking helps leverage contacts to achieve varied business objectives. Understanding and expanding these networks assists in navigating challenges more effectively and capitalising on opportunities for growth and development. Complementary tourism businesses often find it easier to collaborate as they can add value to tourism products without direct competition. However, even competing businesses can collaborate on common issues and opportunities. Despite the benefits of collaborations, many tourism businesses do not collaborate or participate in organisations; this fact is often overlooked by tourism policies becoming overly ambitious in delivering development programmes. This chapter underscores the importance of strategic alignment, proactive engagement, and mutual collaboration in achieving long-term success and resilience in tourism.

## References

Ahn, T. K., & Ostrom, E. (2008). Social capital and collective action. In D. Castiglione, J. W. Van Deth, & G. Wolleb (Eds.), *The handbook of social capital* (pp. 70–100). Oxford University Press.

Ashari, H. A., Heidari, M., & Parvaresh, S. (2014). Improving SMTEs' business performance through strategic use of information communication technology: ICT and tourism challenges and opportunities. *International Journal of Academic Research in Accounting, Finance and Management Sciences*, *4*(3), 1–20.

Ateljevic, J., & Doorne, S. (2004). Diseconomies of scale: A study of development constraints in small tourism girms in central New Zealand. *Tourism and Hospitality Research*, *5*(1), 5–24. 10.1057/palgrave.thr.6040002

Bengtsson, M., & Johansson, M. (2014). Managing coopetition to create opportunities for small firms. *International Small Business Journal*, *32*(4), 401–427. 10.1177/0266242612461288

Bengtsson, M., & Kock, S. (2000). "Coopetition" in business networks – To cooperate and compete simultaneously. *Industrial Marketing Management*, *29*(5), 411–426. 10.1016/s0019-8501(99)00067-x

Bengtsson, M., & Kock, S. (2014). Coopetition – Quo vadis? Past accomplishments and future challenges. *Industrial Marketing Management*, *43*(2), 180–188. 10.1016/j.indmarman.2014.02.015

Binder, P. (2019). A network perspective on organizational learning research in tourism and hospitality. *International Journal of Contemporary Hospitality Management*, *31*(7), 2602–2625. 10.1108/IJCHM-04-2017-0240

Booyens, I., & Rogerson, C. M. (2017). Networking and learning for tourism innovation: Evidence from the Western Cape. *Tourism Geographies*, *19*(3), 340–361. 10.1080/14616688.2016.1183142

Bouncken, R. B., Gast, J., Kraus, S., & Bogers, M. (2015). Coopetition: A systematic review, synthesis, and future research directions. *Review of Managerial Science*, *9*(3), 577–601. 10.1007/s11846-015-0168-6

Bouncken, R. B., & Kraus, S. (2013). Innovation in knowledge-intensive industries: The double-edged sword of coopetition. *Journal of Business Research, 66*(10), 2060–2070. 10.1016/j.jbusres.2013.02.032

Bowie, D., & Buttle, F. (2011). *Hospitality marketing.* Routledge. 10.4324/9780080967929

Buhalis, D., & Leung, R. (2018). Smart hospitality – Interconnectivity and interoperability towards an ecosystem. *International Journal of Hospitality Management, 71*, 41–50. 10.1016/j.ijhm.2017.11.011

Buhalis, D., O'Connor, P., & Leung, R. (2023). Smart hospitality: From smart cities and smart tourism towards agile business ecosystems in networked destinations. *International Journal of Contemporary Hospitality Management, 35*(1), 369–393. 10.1108/IJCHM-04-2022-0497

Camilleri, M. A. (2018). *Travel marketing, tourism economics and the airline product: An introduction to theory and practice.* Springer International Publishing. 10.1007/978-3-319-49849-2

Chin, W. L., Haddock-Fraser, J., & Hampton, M. P. (2017). Destination competitiveness: Evidence from Bali. *Current Issues in Tourism, 20*(12), 1265–1289. 10.1080/13683500.2015.1111315

Cooke, P., & Morgan, K. (1998). *The associational economy: Firms, regions, and innovation.* Oxford University Press.

Crick, J. M., Friske, W., & Morgan, T. A. (2024). The relationship between coopetition strategies and company performance under different levels of competitive intensity, market dynamism, and technological turbulence. *Industrial Marketing Management, 118*, 56–77. 10.1016/j.indmarman.2024.02.005

Curry, O., & Dunbar, R. I. M. (2013). Do birds of a feather flock together? *Human Nature, 24*(3), 336–347. 10.1007/s12110-013-9174-z

Czakon, W., & Czernek, K. (2016). The role of trust-building mechanisms in entering into network coopetition: The case of tourism networks in Poland. *Industrial Marketing Management, 57*, 64–74. 10.1016/j.indmarman.2016.05.010

Damayanti, M., Scott, N., & Ruhanen, L. (2017). Coopetitive behaviours in an informal tourism economy. *Annals of Tourism Research, 65*, 25–35. 10.1016/j.annals.2017.04.007

Della Corte, V., & Aria, M. (2016). Coopetition and sustainable competitive advantage. The case of tourist destinations. *Tourism Management, 54*, 524–540. 10.1016/j.tourman.2015.12.009

Devece, C., Ribeiro-Soriano, D. E., & Palacios-Marqués, D. (2019). Coopetition as the new trend in inter-firm alliances: Literature review and research patterns. *Review of Managerial Science, 13*(2), 207–226. 10.1007/s11846-017-0245-0

Dunbar, R. I. M. (1993). Coevolution of neocortical size, group size and language in humans. *Behavioral and Brain Sciences, 16*(4), 681–694. 10.1017/S0140525X00032325

Dunbar, R. I. M. (1998). The social brain hypothesis. *Evolutionary Anthropology: Issues, News, and Reviews, 6*(5), 178–190. 10.1002/(SICI)1520-6505(1998)6:5<178::AID-EVAN5>3.0.CO;2-8

Erkuş-Öztürk, H. (2011). Emerging importance of institutional capacity for the growth of tourism clusters: The case of Antalya. *European Planning Studies, 19*(10), 1735–1753. 10.1080/09654313.2011.614384

Felzensztein, C., Gimmon, E., & Deans, K. R. (2018). Coopetition in regional clusters: Keep calm and expect unexpected changes. *Industrial Marketing Management, 69*, 116–124. 10.1016/j.indmarman.2018.01.013

Fyall, A., & Garrod, B. (2005). *Tourism marketing: A collaborative approach*. Channel View Publications.

Fyall, A., Garrod, B., & Wang, Y. (2012). Destination collaboration: A critical review of theoretical approaches to a multi-dimensional phenomenon. *Journal of Destination Marketing & Management, 1*(1–2), 10–26. 10.1016/j.jdmm.2012.10.002

Garrod, B., & Fyall, A. (2017). Collaborative destination marketing at the local level: Benefits bundling and the changing role of the local tourism association. *Current Issues in Tourism, 20*(7), 668–690. 10.1080/13683500.2016.1165657

Getz, D., & Carlsen, J. (2000). Characteristics and goals of family and owner-operated businesses in the rural tourism and hospitality sectors. *Tourism Management, 21*(6), 547–560. 10.1016/S0261-5177(00)00004-2

Gilding, M., Brennecke, J., Bunton, V., Lusher, D., Molloy, P. L., & Codoreanu, A. (2020). Network failure: Biotechnology firms, clusters and collaborations far from the world superclusters. *Research Policy, 49*(2). 10.1016/j.respol.2019.103902

Giuseppe, M., Scott, M., Marcello, A., & Giacomo, D. C. (2022). Collaboration and learning processes in value co-creation: A destination perspective. *Journal of Travel Research*. 10.1177/00472875211070349

Gnyawali, D. R., & Park, B.-J. (2009). Co-opetition and technological innovation in small and medium-sized enterprises: A multilevel conceptual model. *Journal of Small Business Management, 47*(3), 308–330. 10.1111/j.1540-627X.2009.00273.x

Gnyawali, D. R., & Park, B.-J. (2011). Co-opetition between giants: Collaboration with competitors for technological innovation. *Research Policy, 40*(5), 650–663. 10.1016/j.respol.2011.01.009

Grauslund, D., & Hammershøy, A. (2021). Patterns of network coopetition in a merged tourism destination. *Scandinavian Journal of Hospitality and Tourism, 21*(2), 192–211. 10.1080/15022250.2021.1877192

Håkansson, H., & Snehota, I. (1995, September). *The burden of relationships or who's next. IMP Conference (11th), Manchester.*

Hall, C. M. (2004). Small firms and wine and food tourism in New Zealand: Issues of collaboration, clusters and lifestyles. In R. Thomas (Ed.), *Small Firms In Tourism* (pp. 167–181). Taylor & Francis.

Hall, C. M., & Williams, A. M. (2020). *Tourism and innovation*. Routledge.

Hill, R. A., & Dunbar, R. I. M. (2003). Social network size in humans. *Human Nature, 14*(1), 53–72. 10.1007/s12110-003-1016-y

Hoholm, T. (2015). Interaction avoidance in networks. *IMP Journal, 9*(2), 117–135. 10.1108/IMP-03-2015-0011

Ingley, C. (2008). *The cluster concept: Cooperative networks and replicability*. Paper presented at the *Conference 'International Council for Small Businesses'*. Naples, Italy.

Kelliher, F., Reinl, L., Johnson, T. G., & Joppe, M. (2018). The role of trust in building rural tourism micro firm network engagement: A multi-case study. *Tourism Management*, *68*, 1–12. 10.1016/j.tourman. 2018.02.014

Kim, N., & Shim, C. (2018). Social capital, knowledge sharing and innovation of small- and medium-sized enterprises in a tourism cluster. *International Journal of Contemporary Hospitality Management*, *30*(6), 2417–2437. 10.1108/ijchm-07-2016-0392

Kozak, M., & Rimmington, M. (1998). Benchmarking: Destination attractiveness and small hospitality business performance. *International Journal of Contemporary Hospitality Management*, *10*(5), 184–188. 10.1108/09596119810227767

Kylänen, M., & Mariani, M. (2012). Unpacking the temporal dimension of coopetition in tourism destinations: Evidence from Finnish and Italian theme parks. *Anatolia*, *23*(1), 61–74. 10.1080/13032917. 2011.653632

Kylänen, M., & Rusko, R. (2011). Unintentional coopetition in the service industries: The case of Pyhä-Luosto tourism destination in the Finnish Lapland. *European Management Journal*, *29*(3), 193–205. 10.1016/j. emj.2010.10.006

Lechner, C., & Leyronas, C. (2012). The competitive advantage of cluster firms: The priority of regional network position over extra-regional networks – a study of a French high-tech cluster. *Entrepreneurship & Regional Development*, *24*(5/6), 457–473. 10.1080/08985626.2011. 617785

Leick, B., & Gretzinger, S. (2020). Business networking in organisationally thin regions: A case study on network brokers, SMEs and knowledge-sharing. *Journal of Small Business and Enterprise Development*, *27*(5), 839–861. 10.1108/JSBED-12-2019-0393

Leiper, N. (1990). Partial industrialization of tourism systems. *Annals of Tourism Research*, *17*(4), 600–605. 10.1016/0160-7383(90)90030-U

Leiper, N., Stear, L., Hing, N., & Firth, T. (2008). Partial industrialisation in tourism: A new model. *Current Issues in Tourism*, *11*(3), 207–235. 10.2167/cit356.0

Levy, M., Loebbecke, C., & Powell, P. (2017). SMEs, co-opetition and knowledge sharing: The role of information systems. *European Journal of Information Systems*, *12*(1), 3–17. 10.1057/palgrave.ejis.3000439

Lim, S., Ok, C. M., & Yang, Y. (2024). A meta-analytic investigation of innovation predictors in tourism and hospitality organizations. *Tourism Management*, *105*. 10.1016/j.tourman.2024.104965

Lin, N. (2008). A network theory of social capital. In D. Castiglione, J. W. Van Deth, & G. Wolleb (Eds.), *The handbook of social capital*. Oxford University Press.

Liu, G., Aroean, L., & Ko, W. W. (2023). Service innovation in business ecosystem: The roles of shared goals, coopetition, and interfirm power. *International Journal of Production Economics*, *255*. 10.1016/j. ijpe.2022.108709

Locke, E. A., & Latham, G. P. (1990). *A theory of goal setting & task performance*. Prentice-Hall.

Locke, E. A., & Latham, G. P. (2017). *New developments in goal setting and task performance* (First issued in paperback ed.). Routledge. (Original work published 2013).

Marasco, A., De Martino, M., Magnotti, F., & Morvillo, A. (2018). Collaborative innovation in tourism and hospitality: A systematic review of the literature. *International Journal of Contemporary Hospitality Management, 30*(6), 2364–2395. 10.1108/ijchm-01-2018-0043

McComb, E. J., Boyd, S., & Boluk, K. (2017). Stakeholder collaboration: A means to the success of rural tourism destinations? A critical evaluation of the existence of stakeholder collaboration within the Mournes, Northern Ireland. *Tourism and Hospitality Research, 17*(3), 286–297. 10.1177/1467358415583738

McGrath, H., O'Toole, T., & Canning, L. (2019). Coopetition: A fundamental feature of entrepreneurial firms' collaborative dynamics. *The Journal of business & Industrial Marketing, 34*(7), 1555–1569. 10.1108/jbim-10-2018-0287

McTiernan, C., Musgrave, J., & Cooper, C. (2021). Conceptualising trust as a mediator of pro-environmental tacit knowledge transfer in small and medium sized tourism enterprises. *Journal of Sustainable Tourism, 31(4)*, 1014–1031. 10.1080/09669582.2021.1942479

Michael, E. J. (2007a). *Micro-clusters and networks: The growth of tourism*. Routledge. 10.4324/9780080464909

Michael, E. J. (2007c). Micro-clusters: Antiques, retailing and business practice. In E. J. Michael (Ed.), *Micro-clusters and networks: The growth of tourism* (pp. 63–78). Routledge.

Moore, J. F. (1993). Predators and prey: A new ecology of competition. *Harvard Business Review, 71*(3), 75–86.

Moretti, A., Martini Barzolai, M., & Cutugno, M. (2024). Ambidexterity of hospitality trade associations: A social capital perspective for destination coordination and cooperation dynamics. *Current Issues in Tourism*. 10.1080/13683500.2024.2414929

Morrison, A., Lynch, P., & Johns, N. (2004). International tourism networks. *International Journal of Contemporary Hospitality Management, 16*(3), 197–202. 10.1108/09596110410531195

Nalebuff, B. J., & Brandenburger, A. M. (1997). Co-opetition: Competitive and cooperative business strategies for the digital economy. *Strategy & Leadership, 25*(6), 28–33. 10.1108/eb054655

Nguyen, T. Q. T., Johnson, P., & Young, T. (2022). Networking, coopetition and sustainability of tourism destinations. *Journal of Hospitality and Tourism Management*. 10.1016/j.jhtm.2022.01.003

Nguyen, T. Q. T., Nguyen, V. T., Hoang, T. T. H., Tran, T. H. T., & Nguyen, T. P. T. (2024). Social networking, environmental awareness and sustainable tourism development in Da Nang, Vietnam. *Tourism and Hospitality Research*. 10.1177/14673584241234269

Nishimura, J., & Okamuro, H. (2011). Subsidy and networking: The effects of direct and indirect support programs of the cluster policy. *Research Policy, 40*(5), 714–727. 10.1016/j.respol.2011.01.011

Okumus, F., Altinay, L., & Chathoth, P. (2010). *Strategic management in the international hospitality and tourism industry*. Routledge. 10.4324/9780080940465

Pechlaner, H., & Tschurtschenthaler, P. (2003). Tourism policy, tourism organisations and change management in alpine regions and destinations: A European perspective. *Current Issues in Tourism, 6*(6), 508–539. 10.1080/13683500308667967

Pechlaner, H., & Volgger, M. (2012). How to promote cooperation in the hospitality industry. *International Journal of Contemporary Hospitality Management, 24*(6), 925–945. 10.1108/09596111211247245

Perkins, R., Khoo-Lattimore, C., & Arcodia, C. (2020). Understanding the contribution of stakeholder collaboration towards regional destination branding: A systematic narrative literature review. *Journal of Hospitality and Tourism Management, 43*, 250–258. 10.1016/j.jhtm.2020.04.008

Perles-Ribes, J., Rodríguez-Sánchez, I., & Ramón Rodríguez, A. (2014). Innovative tourism clusters: Myth or reality? Empirical evidence from Benidorm. 10.2139/ssrn.2424737

Perry, M. (2001). *Shared trust in New Zealand: Strategies for a small industrial country*. Institute of Policy Studies, Victoria University of Wellington.

Perry, M. (2007). From networks to clusters and back again: A decade of unsatisfied policy aspiration in New Zealand. In R. MacGregor & A. Hodgkinson (Eds.), *Small business clustering technologies: Applications in marketing, management, IT and economics* (pp. 160–183). IGI Global.

Petrou, A., Pantziou, E. F., Dimara, E., & Skuras, D. (2007). Resources and activities complementarities: The role of business networks in the provision of integrated rural tourism. *Tourism Geographies, 9*(4), 421–440. 10.1080/14616680701647634

Pforr, C. (2004). Policy-making for sustainable tourism. In F. D. Pineda & C. A. Brebbia (Eds.), *Sustainable tourism* (pp. 83–94). WIT Press.

Phillipson, J., Gorton, M., & Laschewski, L. (2006). Local business cooperation and the dilemmas of collective action: Rural micro-business networks in the North of England. *Sociologia Ruralis, 46*(1), 40–60. 10.1111/j.1467-9523.2006.00401.x

Pilving, T., Kull, T., Suškevics, M., & Viira, A. H. (2022). Diverse networks in regional tourism: Ties that foster and hinder the development of rural-urban tourism collaboration in estonia. *European Journal of Tourism Research, 30*. 10.54055/ejtr.v30i.2596

Porto-Gomez, I., Aguirre-Larracoechea, U., & Zabala-Iturriagagoitia, J. M. (2018). Tacit coopetition: Chimera or reality? Evidence from the Basque Country. *European Planning Studies, 26*(3), 611–634. 10.1080/09654313.2017.1402866

Quaranta, G., Citro, E., & Salvia, R. (2016). Economic and social sustainable synergies to promote innovations in rural tourism and local development. *Sustainability, 8*(7), 668. 10.3390/su8070668

Raisi, H., Baggio, R., Barratt-Pugh, L., & Willson, G. (2020). A network perspective of knowledge transfer in tourism. *Annals of Tourism Research, 80*, 102817. 10.1016/j.annals.2019.102817

Raza-Ullah, T., & Kostis, A. (2020). Do trust and distrust in coopetition matter to performance? *European Management Journal, 38*(3), 367–376. 10.1016/j.emj.2019.10.004

Roberts, S. G. B., Dunbar, R. I. M., Pollet, T. V., & Kuppens, T. (2009). Exploring variation in active network size: Constraints and ego characteristics. *Social Networks, 31*(2), 138–146. 10.1016/j.socnet.2008.12.002

Rodríguez, I., Williams, A. M., & Hall, C. M. (2014). Tourism innovation policy: Implementation and outcomes. *Annals of Tourism Research, 49*, 76–93. 10.1016/j.annals.2014.08.004

Ruokolainen, J., & Igel, B. (2022). The elusiveness of business networks – Why do science park firm tenants not collaborate with neighbors? *Industrial Marketing Management, 101*, 113–124. 10.1016/j.indmarman. 2021.11.011

Rusko, R. (2014). Mapping the perspectives of coopetition and technology-based strategic networks: A case of smartphones. *Industrial Marketing Management, 43*(5), 801–812. 10.1016/j.indmarman.2014.04.013

Rusko, R. (2024). Coopetition networks in tourism destinations: A literature review. In M. A. Camilleri (Ed.), *Tourism planning and destination marketing*, 2nd Edition (pp. 79–92). Emerald Publishing. 10.1108/978-1-80455-888-120241004

Safonov, A., Hall, C. M., & Prayag, G. (2023). Non-collaborative behaviour of accommodation businesses in the associational tourism economy. *Journal of Hospitality and Tourism Management, 54*, 98–107. 10.1016/j.jhtm.2022.12.007

Saufi, A., O'Brien, D., & Wilkins, H. (2014). Inhibitors to host community participation in sustainable tourism development in developing countries. *Journal of Sustainable Tourism, 22*(5), 801–820. 10.1080/ 09669582.2013.861468

Scaringella, L., & Radziwon, A. (2018). Innovation, entrepreneurial, knowledge, and business ecosystems: Old wine in new bottles? *Technological Forecasting and Social Change, 136*, 59–87. 10.1016/j. techfore.2017.09.023

Schiavone, F., & Simoni, M. (2011). An experience-based view of coopetition in R&D networks. *European Journal of Innovation Management, 14*(2), 136–154. 10.1108/14601061111124867

Schrank, A., & Whitford, J. (2011). The anatomy of network failure. *Sociological Theory, 29*(3), 151–177. 10.1111/j.1467-9558.2011.01392.x

Scott, D., Gössling, S., & Hall, C. M. (2012). International tourism and climate change. *Wiley Interdisciplinary Reviews: Climate Change, 3*(3), 213–232. 10.1002/wcc.165

Slattery, P. (2002). Finding the hospitality industry. *Journal of Hospitality, Leisure, Sport & Tourism Education, 1*(1), 19–28.

Smith, R. A., & Henderson, J. C. (2008). Integrated beach resorts, informal tourism commerce and the 2004 tsunami: Laguna Phuket in Thailand. *The International Journal of Tourism Research, 10*(3), 271–282. 10.1002/jtr.659

Son, I. S., Huang, S., & Padovan, D. (2021). Realising the goals of event leveraging: The tourism and hospitality SME perspective. *Journal of*

*Hospitality and Tourism Management, 49*, 253–259. 10.1016/j.jhtm.2021.09.018

Song, H. (2011). *Tourism supply chain management.* Routledge. 10.4324/9780203804391

Sørensen, F. (2007). The geographies of social networks and innovation in tourism. *Tourism Geographies, 9*(1), 22–48. 10.1080/14616680601092857

Steinbruch, F. K., Nascimento, L. d. S., & de Menezes, D. C. (2022). The role of trust in innovation ecosystems. *Journal of Business & Industrial Marketing, 37*(1), 195–208. 10.1108/JBIM-08-2020-0395

Taylor, P., McRae-Williams, P., & Lowe, J. (2007). The determinants of cluster activities in the Australian wine and tourism industries. *Tourism Economics, 13*(4), 639–656. 10.5367/000000007782696050

Tinsley, R., & Lynch, P. A. (2008). Differentiation and tourism destination development: Small business success in a close-knit community. *Tourism and Hospitality Research, 8*(3), 161–177. 10.1057/thr.2008.26

Tunisini, A., & Marchiori, M. (2020). Why do network organizations fail? *Journal of Business & Industrial Marketing, 35*(6), 1011–1021. 10.1108/JBIM-01-2019-0056

van der Zee, E., & Vanneste, D. (2015). Tourism networks unravelled; a review of the literature on networks in tourism management studies. *Tourism Management Perspectives, 15*, 46–56. 10.1016/j.tmp.2015.03.006

Volgger, M., & Pechlaner, H. (2014). Requirements for destination management organizations in destination governance: Understanding DMO success. *Tourism Management, 41*, 64–75. 10.1016/j.tourman.2013.09.001

Volgger, M., & Pechlaner, H. (2015). Governing networks in tourism: What have we achieved, what is still to be done and learned? *Tourism Review, 70*(4), 298–312. 10.1108/TR-04-2015-0013

von Friedrichs Grängsjö, Y. (2003). Destination networking: Co-opetition in peripheral surroundings. *International Journal of Physical Distribution and Logistics Management, 33*(5), 427–448. 10.1108/09600030310481997

Wang, Y., & Krakover, S. (2008). Destination marketing: Competition, cooperation or coopetition? *International Journal of Contemporary Hospitality Management, 20*(2), 126–141. 10.1108/09596110810852122

Webb, T., Beldona, S., Schwartz, Z., & Bianco, S. (2021). Growing the pie: An examination of coopetition benefits in the US lodging industry. *International Journal of Contemporary Hospitality Management, 33*(12), 4355–4372 10.1108/IJCHM-03-2021-0340

Werner, K., Dickson, G., & Hyde, K. (2015). Coopetition and knowledge transfer dynamics: New Zealand's regional tourism organizations and the 2011 Rugby World Cup. *Event Management, 19*. 10.3727/152599515X14386220874841

Williams, P. W., & Ponsford, I. F. (2009). Confronting tourism's environmental paradox: Transitioning for sustainable tourism. *Futures, 41*(6), 396–404. 10.1016/j.futures.2008.11.019

Yu, X., Kim, N., Chen, C., & Schwartz, Z. (2012). Are you a tourist? Tourism definition from the tourist perspective. *Tourism Analysis, 17*(4), 445–457. 10.3727/108354212X13473157390687

Zach, F., & Racherla, P. (2011). Assessing the value of collaborations in tourism networks: A case study of Elkhart County, Indiana. *Journal of Travel & Tourism Marketing, 28*(1), 97–110. 10.1080/10548408.2011.535446

Zach, F. J., & Hill, T. L. (2017). Network, knowledge and relationship impacts on innovation in tourism destinations. *Tourism Management, 62*, 196–207. 10.1016/j.tourman.2017.04.001

Zhang, X., Song, H., & Huang, G. Q. (2009). Tourism supply chain management: A new research agenda. *Tourism Management, 30*(3), 345–358. 10.1016/j.tourman.2008.12.010

Zhou, W.-X., Sornette, D., Hill, R. A., & Dunbar, R. I. M. (2005). Discrete hierarchical organization of social group sizes. *Proceedings of the Royal Society B: Biological Sciences, 272*(1561), 439–444. 10.1098/rspb.2004.2970

# 4 The complexities of collaboration at destinations

## Introduction

Collaboration at the destination level is inherently complex due to the various actors involved in tourism, including businesses, organisations, governments, and tourists. While collaboration with tourism organisations is often promoted as advantageous for business competitiveness and offers various benefits, there are instances where networks fail, and businesses may choose not to engage (Schrank & Whitford, 2011; Tunisini & Marchiori, 2020). Many network organisations, such as tourism bodies, provide marketing services, business-related information, and advocacy, primarily for a membership fee. While businesses can participate in these networks, they may not actively contribute to them, as the exchange is often based on monetary conditions and can diminish in value over time (Phillipson et al., 2006). As more participants engage in tourism collaboration, managing, coordinating, and controlling these interactions become increasingly challenging. Collaboration is an emergent property that evolves, changes, and transforms both within itself and within the broader collaboration domain through interactions with external elements and the receipt of feedback (Gray, 1989). Participants' worldviews and perceptions also shift during this process with increased knowledge and experience. Consequently, given the dynamic nature of participant perspectives, predicting the effectiveness of collaboration becomes increasingly challenging.

## Reductionism and complexity

Tourism and hospitality research frequently employs a variety of theoretical approaches to explore different aspects of complex issues, reinforcing the argument that tourism lacks a robust standalone theory (Farrell & Twining-Ward, 2004; Faulkner & Russell, 2003). The complexity of tourism, which spans space and time, often results in research outcomes that are specific to particular contexts and temporal conditions (Kesić, 2016).

DOI: 10.4324/9781003293606-4

Even concepts such as networks, clusters, or ecosystems alone cannot fully explain tourism phenomena. Research often focuses on isolated components of the system that can help address specific issues and neglects the overall functioning of the entire system. Consequently, a reductionist approach prevails in tourism research and practice (Farrell & Twining-Ward, 2004).

Reductionism involves breaking down complex phenomena into smaller, manageable parts, under the assumption that a detailed focus on these individual components allows for the exploration of sophisticated issues such as human behaviour, society, and the world (Zahra & Ryan, 2007). In tourism, this reductionist perspective attempts to explain certain elements of the system; however, it falls short in addressing the effects of the system as a whole. According to reductionist views, events can be predicted based on initial conditions (Faulkner & Russell, 2003). However, fragmenting a complex phenomenon into parts provides a simplified analysis of evident elements (Tronvoll et al., 2018) that often overlooks the emergent properties of complex environments (Kesić, 2016). These limitations have led to calls to transcend reductionism in tourism, as it inadequately explains non-linear behaviour by oversimplifying complex problems (Barile et al., 2018; Faulkner & Russell, 2003).

Complexity arises when the components of a system interact in a non-linear manner. As the system evolves, it may increase in complexity until the next self-organisation process occurs (Baggio, 2008). Such complex systems are real and not merely theoretical constructs (Farrell & Twining-Ward, 2004). Feedback loops within the system allow for learning and evolution over time and space. Emergent properties of these systems are neither directly identifiable nor predictable based solely on the knowledge of individual components and their interactions (Chapman, 2009). From a systemic perspective, tourism lacks clear boundaries and rules. As a social system, tourism involves subjective interpretations of behaviour, underscoring the necessity of understanding it from systemic and holistic viewpoints. Thus, although the reductionist approach prevails in tourism research and practice, there is a growing recognition that tourism embodies complex systems that cannot be deconstructed to predict overall system behaviour.

Tourism, fundamentally based on social systems representing behavioural phenomena, requires a shift beyond a reductionist approach (Faulkner & Russell, 2003; Leiper, 1992). Tourism systems are interdependent, non-deterministic, and non-linear, encompassing tourists, organisations, and broader environments (Farrell & Twining-Ward, 2004). Consequently, predicting system behaviour through a reductionist lens proves challenging. Furthermore, tourism system models, such as those introduced by Leiper or Mill and Morrison (Leiper, 1979; Mill & Morrison, 1985), fail to account for potential disruptions or unpredictable events. While these models include external environments, they visually represent tourism systems as static and in an established state. As schematic representations of complex phenomena, these models do not capture the non-deterministic behaviour inherent in tourism systems. This limitation is critical as tourism planners and managers may mistakenly believe they can predict and control full dynamics in tourism. Leiper's 'basic tourism system' in fact consists of interdependent tourism systems interacting simultaneously, and thus serves as a useful representation for understanding and developing systems thinking. A systemic framework enables a holistic perspective, allowing for the interpretation of interactions across different scales (Polese, 2018). Given the limitations of reductionism in explaining tourism as a complex phenomenon, a holistic approach offers a better opportunity to understand changing dynamics (Baggio, 2008). This systemic perspective is essential, as it addresses the social nature of interactions within the tourism system and the complexity of challenges such as pollution, climate change, sustainability, resilience, and collaboration. For instance, sustainability issues require a holistic perspective to achieve transformative shifts in research and practice, as economic aspects often overshadow social and environmental concerns (Barile & Saviano, 2018).

While reductionist and holistic perspectives may seem opposed, they are not mutually exclusive. Reductionism dissects complex phenomena into components, while a holistic approach examines the system as a whole. While the argument for adopting a holistic perspective stems from the limitations of simplified methods in explaining tourism behaviour, the choice of perspective often reflects the observer's beliefs. Consequently, the description, interpretation, and exploration of tourism will reflect the epistemological views of researchers, practitioners, marketers, and managers. Therefore, tourism, as a social phenomenon, is filled with subjective meanings and interpretations. This often leads to the application of lay theories, which provide a sense of explanation and control over social reality (Furnham, 1988). While lay theories can occasionally offer insights into the social world, they contrast with scientific methods and reflect personal interpretations of reality. This form of apophenia is intrinsic to tourism planning, development, and management within the collaboration domain. While reductionism has its place, a holistic perspective is essential for understanding the complexities of tourism systems. The dichotomy between reductionist and holistic approaches creates a paradox in tourism management. On the one hand, the reductionist approach struggles to predict and describe the emergent properties of tourism systems (Baggio et al., 2010; Farrell & Twining-Ward, 2004). On the other hand, the holistic approach is often too broad and complex to be effectively applied within a destination context (Speakman, 2017). Achieving a balance between these perspectives is crucial for effective tourism development. The complexity of tourism systems involves understanding several critical aspects.

## Self-organisation and complex system behaviour

A key intrinsic characteristic of complex systems is their self-organising nature. From an ecological perspective, particularly of animal behaviour, studies provide examples of self-organisation such as flocks of birds or schools of fish (Baggio et al., 2010; Scott et al., 2008). However, whether these animals consciously self-organise or possess an understanding of their actions raises significant questions when translating these behaviours to human interactions within tourism. Human behaviour is inherently difficult to predict, which adds to the complexity of tourism systems. For instance, self-interest theory often falls short in accurately predicting business-to-business behaviour on smaller scales, and becomes even less reliable at larger scales. If human behaviour was primarily driven by basic instincts and needs, similar to less evolved species, predicting such behaviour might be simpler. Thus, acknowledging the role of consciousness is crucial when using these animal behaviours as analogies for human systems. Although behaviourism traditionally focused on observable and measurable actions, often neglecting consciousness (Bush, 2006), a deeper understanding of complex behaviours may require recognising the importance of consciousness (Mills, 1998). Our observations and descriptions of these behaviours are products of cognitive processes to measure and explain various phenomena, including tourism systems. As tourists are at the core of tourism systems, comprehending their behaviour within the framework of complex systems thinking is essential.

## The space-time dimension in tourism

The significance of the space-time dimension in tourism highlights the complexities inherent in tourism systems. Seemingly inconsequential events can trigger unpredictable and wide-ranging consequences. For instance, the metaphor of a butterfly flapping its wings leading to a cyclone illustrates how an initial event can set off a chain reaction with significant outcomes (Gleick, 1987; Lorenz, 1963, 1993). While the butterfly does not directly cause the cyclone, the interconnectedness of subsequent events leads to the cyclone, even if the relationship between the initial input and the outcome is not immediately obvious. The challenge of tracing these events in a linear sequence is further complicated by their spread across space and time. Observing and understanding these occurrences within a human lifespan is difficult, as one can only witness either the butterfly or the cyclone at a given moment. The increasing number of interactions, causal relationships, feedback loops, emergent properties, and bifurcation points contribute to systemic behaviour, making observation and prediction challenging. Consequently, the system's complexity arises from its comprehensive and often unobservable nature, which cannot be fully understood by examining individual components in isolation.

Even if reductionism could explain complex systems by analysing smaller elements or extrapolating local behaviours, the vast number of transactions, decisions, and simultaneous conscious and unconscious actions underpinning these systems presents a significant challenge. Large complex systems become unpredictable and difficult to fully comprehend. While a system at equilibrium may be observable, it cannot be entirely understood, as any individual observable component loses significance when viewed from a broader space-time perspective.

### Metaphors of ecosystems in business and tourism

In business and tourism literature, the complexity of ecosystems is often used as a metaphor to draw analogies to natural systems. This comparison is figurative rather than literal and aims to illustrate that there is something more substantial to competitiveness, resilience, and sustainability than simply the combination of individual elements. The concept of a business ecosystem originally described large companies as examples of such systems (e.g., Moore, 1993), where users interact through the company's software and hardware products as part of an 'ecosystem'. The objective is to enhance profitability by creating a stable environment that encourages repeat purchases of products and services throughout a user's lifetime. However, translating the properties of ecological systems into business-to-business relationships poses challenges due to differences in the nature and dynamics of interactions. The concise and social aspects of business interactions do not always align with the dynamics of ecological systems, making direct comparisons difficult.

In this context, a destination can be viewed as a 'large' company striving to create an 'ecosystem' for tourists, aiming to maximize their length of stay and depth of travel within a region. Tourism systems operate effectively and become cohesive only when tourists are present, making it essential to understand their behaviour to design successful destinations. Rather than focusing exclusively on marketing and promoting large businesses, the emphasis should be on creating immersive experiences and fostering lasting connections with locations. This approach can cultivate a sustainable economy that shapes tourist behaviour and influences their future decisions.

## Considerations for tourism collaboration governance

Tourism governance typically operates within a multilevel framework, with various bodies functioning at different geographical levels. The autonomy of decision-making at each level, however, is dependent upon a country's governance system. Tourism organisations commonly focus on marketing a country and its destinations, yet management concerns often remain ambiguous. This emphasis on marketing underscores the need to understand the role of tourism within the broader system and at higher levels to address the challenges of tourism development comprehensively. Consequently, it is crucial for each level of governance to understand system properties from its unique perspective. In practical terms, it is essential to recognise that what is measured dictates subsequent actions (Stiglitz, 2009). Therefore, short-term initiatives must be considered alongside their long-term consequences on development trajectories, so that initiatives are designed with a future-oriented approach, ensuring they align with long-term sustainable goals and can be effectively worked backward from desired outcomes. While destination management plans are important for tourism development, they often overemphasise indicators such as tourist numbers, nights, and expenditure. This focus tends to overlook the long-term aspects of tourism development, failing to address the comprehensive nature of tourism. Consequently, destinations become growth-dependent, relying on strategies that are ultimately unsustainable or introduced repeatedly with similar content after predefined timeframes lapse. Although development plans may incorporate environmental and social aspects, these indicators are often undervalued due to measurement challenges, leading to an overemphasis on economic factors (Barile & Saviano, 2018; Wijesinghe, 2022). As a result, tourism development strategies tend to be repetitive, with economic growth remaining the primary success indicator.

### *Opportunities and threats of tourism development*

Tourism development presents both opportunities and threats to local communities. Consequently, prioritising the commercialisation of tourism organisations' functions and solely marketing tourism businesses should not be the primary focus. For instance, the distinction between regional tourism organisations (RTOs) and industrial associations often becomes blurred. RTOs frequently charge fees for membership to market tourism businesses, yet hospitality businesses such as restaurants or motels may not always be included. Moreover, many businesses now manage their digital marketing independently. Government tourism organisations often operate with power groups or large companies instead of engaging directly with individual small businesses. Rather than focusing on marketing and membership fees, the emphasis should shift to advancing tourism development through a deeper understanding of various tourism systems, and supporting them or developing new ones. From a regional and/or national perspective, support for local tourism organisations should extend to include broader local development needs. In the current political and economic climate, the role of government in tourism has transitioned from a traditional public administration model, focusing on implementing policies for the public good, to a corporatist model that emphasises efficiency, investment returns, market dynamics, and stakeholder relations (Hall, 1999; Perry, 2007). This shift presents significant policy challenges in tourism, as the trend towards privatising and commercialising functions of government organisations has profoundly affected the nature of national government involvement in tourism.

### Prioritising intra-national tourism systems

Developing and prioritising intra-national tourism systems is essential. Domestic tourism should be prioritised before international tourism, despite the latter appearing more economically justified. Local demand is crucial for the competitiveness of tourism offerings (Enright & Newton, 2004). Unlike international tourists, who visit temporarily, locals reside in the country and can be attracted during shoulder seasons or periods of international disruptions. Various national systems can be seamlessly integrated into international ones, allowing for manageable effects. While Leiper's tourism system provides a simplified model (Leiper, 1979), it assumes the existence of multiple systems. Although it is argued that tourism systems are primarily based on inter-regional frameworks (Leiper, 1992), intra-regional tourism systems should also be recognised and developed.

### The importance of local systems and tourist flows

To effectively manage a destination, it is beneficial to deconstruct it into smaller subsystems. Evidence suggests that local tourism organisations can play a significant role in the management of their destinations (Beaumont & Dredge, 2010). Instead of focusing solely on regional, national, or international scales, it is worthwhile to examine tourism opportunities at a local scale, as seen in the case of micro-clusters (Michael, 2007b; Phillipson et al., 2006; Sigurðardóttir & Steinthorsson, 2018), even if they seem small. The impacts at this level are smaller and more manageable. Moreover, local consumption ensures that money remains inside an area, promoting local production. Given that hierarchical structures already exist within the administrative borders of many countries through various tourism organisations, these should be recognised or repositioned as a multiscale system. This approach allows each level to operate autonomously within the broader system, while emphasising communication flows between levels, whether downward or upward, without imposing top-down directives. Furthermore, when and where tourists travel within a destination must be acknowledged, as managing tourist flows and addressing seasonality are vital aspects beyond marketing. Issues such as tourist distribution across space and time must be recognised and addressed to maintain sustainability in terms of social, environmental, and economic impacts. Therefore, creating local tourism systems within regions should not be overlooked.

At the international level, tourism development should extend beyond marketing efforts. Countries need to consider the larger scale of tourist flows into and within their borders, addressing the management of these flows and the factors that influence them. A key question is understanding how a country, given its location, positions itself within supranational tourism systems. This analysis may reveal opportunities for the development and management of tourist flows into and out of the country. Each level of the system must consider the broader context of tourist flows, focusing on how tourists move into, within, and out of regions or locations. Tourism development occurs in constant interaction with specific places, making it impractical to think beyond the immediate level of each location. Figure 4.1 illustrates types of collaboration, ranging from individual firms to governance, across local, regional, national, and international levels, emphasising the scope and scale of collaborative efforts in tourism systems.

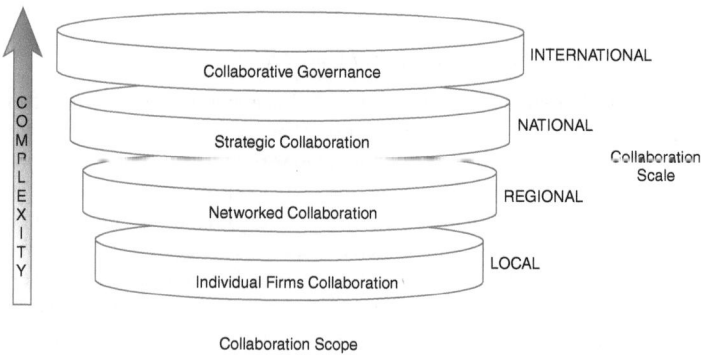

*Figure 4.1* Scope of collaboration across levels.

### The role of geographical context in tourism business behaviour

The context of geographical proximity is significant for understanding tourism business behaviour. Spatial proximity facilitates the consolidation of businesses by creating a community context and a sense of place that allows both formal and informal collaborations to emerge. When businesses are located nearby, they complement each other to satisfy tourist needs, often exhibiting co-opetition behaviour such as, for example, exchanging customers to retain them, pooling staff or resources during busy times, or coordinating their opening hours when demand is low (Chapter 3). It is important to emphasise that collaborative behaviour is often situated within a specific spatial context. On a smaller scale, businesses tend to interact more frequently, leading to a deeper understanding of one another and a wide range of collaborative behaviours beyond basic business exchange (Pilving et al., 2022; Pyke et al., 1990). In larger agglomerations, businesses may become more scattered, resulting in the formation of smaller clusters. For example, businesses in large cities often group into micro-clusters where collaboration occurs within small geographical areas (Arzaghi & Henderson, 2008).

While clusters could be considered on an aggregate level, encompassing entire industries (Porter, 1990), the associated benefits are primarily derived from close-knit communities of co-located businesses. When these benefits depend on direct, personal face-to-face communication, efficient information flow, reciprocity, and informal management structures, clusters larger than approximately 150 individuals (responsible for business strategic behaviour or entrepreneurs) may be less effective in this context (Hill & Dunbar, 2003; Roberts et al., 2009), eventually breaking up into smaller groups. Additionally, in this regard, co-location supports collaboration by reducing transaction cost and increasing interaction frequency (Curry & Dunbar, 2013), promoting reciprocal relationships (Axelrod, 1984/2009). Local cluster size plays a crucial role as it correlates with natural social network structures where communities of 30 to 150 individuals provide more effective structures for sharing and assimilation of information (Hill & Dunbar, 2003; Roberts et al., 2009; Zhou et al., 2005). This suggests that the operational social world is often smaller than it might appear from a policy perspective (Zhou et al., 2005). Although businesses sharing a common market within a specific location tend to be spatially bound, resulting in denser relational dynamics within their area (Garrod & Fyall, 2017), this does not restrict them from collaborating with businesses and organisations outside their immediate locality or even their city or region. This principle also extends to businesses offering services beyond their physical location. For instance, a business with an office in a city might manage activities such as cycling, rafting, or other activity tours outside that city. In such cases, they often collaborate with businesses in the geographical areas where these activities take place. Thus, their operations can be considered spatially co-located, even if their physical presence is not. Recognising geographical proximity as a spatial domain is especially relevant to tourism, where many businesses, such as tour and transport companies, offer services that may be substantially dispersed across space and time.

However, geographical proximity can also lead to tensions (Kallmuenzer et al., 2021; Kelliher et al., 2018). When businesses operate in close proximity, they have more opportunities to observe and adopt each other's behaviour, pricing, and practices (Felzensztein et al., 2018; Kylänen & Rusko, 2011). Consequently, businesses may engage in competitive behaviour to capture a greater market share (Bengtsson & Kock, 2014; Wang & Krakover, 2008). Typically, larger corporate businesses, which possess the resources and capabilities to independently meet tourist needs, may seek monopolistic positions (Perry, 2007). These corporations often receive support from RTOs, which view them as valuable for generating revenue and employment. They have the leverage to lobby for their interests and maintain visibility within tourism organisations. However, this focus on major stakeholders is problematic when considering that about 80% of tourism businesses are micro- to medium-sized (UNWTO, 2023).

Therefore, businesses that are physically or operationally co-located may not only collaborate but may also fiercely compete, especially when they share common resources. It is crucial to consider how distance and physical presence shape collaborative behaviour, as businesses with headquarters located further away may have different priorities for collaboration and may be more competitive. These businesses often lack the same level of dedication to local context and sense of place that local businesses have. Moreover, decision-makers for these businesses may not be physically present in the area, leading to strategies and supply chains managed from a distance. This can result in a lack of contextual interactions and collaboration, even when co-location exists. Thus, whether through physical or operational co-location, businesses may compete for resources and market access, often choosing not to collaborate.

This dynamic partially explains why many cluster initiatives fail (Markusen, 2003), as they tend to focus solely on the benefits of geographical proximity without accounting for the complex nature of collaboration and competition among businesses. Therefore, geographical proximity can facilitate both collaborative and non-collaborative behaviours, helping to determine whether businesses share a personal or professional match. Successful collaboration initiatives require a deeper understanding of each business's motivations and interests, including non-collaborative behaviours and the broader contextual factors that shape the collaborative environment.

### *Rethinking the role of regional tourism organisations*

Tourism is often subsumed under the economic development agenda, and this tends to overlook the complex impacts of tourism. While some argue that creating connections among tourism businesses requires external government interventions (Zhong & Tang, 2018), it is crucial to apply financial incentives cautiously, as they may not be sufficient to sustain long-term collaboration (Nishimura & Okamuro, 2011). Collaboration can eventually fail without shared interests, issues, or goals – the things that bind businesses together (Letaifa & Rabeau, 2013). Therefore, in addition to financial support, it is vital to focus on other factors, such as cultivating shared values and beliefs, to facilitate a clear understanding of business commitments. Such an understanding can help overcome fundamental barriers and promote a collaborative mindset among businesses, creating an ecosystem conducive to collective growth and development.

The emphasis on regional promotion without adequately developing tourism or understanding its dynamics (Pike & Page, 2014) positions RTOs more in a role similar to professional industry associations. Without a clear strategy and objectives for tourism development, the focus shifts primarily to marketing a region (Fyall et al., 2012) and attracting events with questionable economic justification and legacies (Ziakas, 2024), often merely to populate the organisation's website with as many tourism businesses as possible. The passive approach to tourism management turns the RTO into a commercialised entity, judged primarily on the efficiency and effectiveness of its funding (Pechlaner et al., 2011; Pechlaner et al., 2012). From one perspective, it makes sense for tourism organisations to primarily focus on promoting a region. In this context, it might seem inadequate to view a petrol station or retail store as part of tourism environment. However, from another angle, recognising all the touchpoints that tourists encounter within a region shifts the focus from mere marketing to creating immersive ecosystems that provide seamless experiences. Only through a holistic and multifaceted approach can the issue of a collaborative environment be meaningfully addressed, leading to long-term, sustainable development. Given the significant concerns about the impacts of tourism and its long-term sustainability, it is vital to understand tourism as a whole in order to address the potential challenges and opportunities within tourism. A comprehensive vision that understands tourism as an interconnected system is essential for designing appropriate policies. RTOs should play a crucial role in translating this broader understanding and integrating diverse elements to form a cohesive whole from a tourist perspective.

Tourism governance primarily focuses on return on investment, spending efficiency, and the involvement of major stakeholders, relying on them to generate returns. It is assumed that sustainability can be achieved by increasing visitor numbers which translates into personal, business, and regional incomes that, in turn, can address socio-economic challenges (e.g., López-Sanz et al., 2021). A growth mindset prevails, with indicators usually being compared to the previous quarters, years, or periods and performance often evaluated by tourist expenditure, nights, and numbers (Aminudin et al., 2018; Butcher, 2021; Gössling et al., 2021). However, it is hard to define precisely what exactly tourists spend on during their trip. For instance, tourism satellite accounts measure ostensibly tourism-related products and services outputs, which in turn are used to underline the economic benefits of tourism, while leaving well-being, environment, or sustainability impacts out of the scope. First, the designation of businesses as tourism-related is arbitrary. Second, there is no measurement of well-being, environmental impacts, or sustainability. This is why sustainability is usually approached from the economic perspective too.

The measurements we choose ultimately shape our actions and may lead to misguided policy decisions (Stiglitz, 2009). As Stiglitz (2009) notes, government outputs are often measured based on inputs, implying that when a government increases its spending, output appears to rise, regardless of expenditure efficiency. For example, if a government chooses to address environmental concerns, this may potentially lower GDP growth even though such concerns require additional spending (Stiglitz, 2009). Surprisingly, economic growth appears to induce negative impacts while simultaneously providing solutions to these challenges (Butcher, 2021). Without a shift in mindset, proposed solutions often appear to counter the prevailing growth-based economies. Hence, tourism policy should challenge and balance the growth-led paradigm of tourism development in favour of long-term sustainability and resilience (Gössling et al., 2016). The assumption that economic development alone can solve all issues, supposedly serving the public interest, is overly simplistic. Collaboration focused mainly on economic gains often limits opportunities to build the social capital needed to address challenges beyond the capacity of any single firm, such as social and environmental issues. Given this context, an overemphasis on corporatist approaches risks diminishing the impacts of tourism on societies and environments, and undermines the essential role of collaborations in achieving sustainable development goals.

Ideally, associations should align their offerings with the needs of businesses and, in the case of tourism organisations, with long-term development plans that extend beyond merely increasing tourist numbers and their expenditures. This alignment can be particularly challenging when RTOs are subordinate to higher hierarchical structures and must justify their use of taxpayer money. Furthermore, government tourism associations primarily cultivate meaningful collaboration among major stakeholders who hold significant influence over capital flows and thus possess market power (Pforr, 2004; Wesley & Pforr, 2010). Also, a significant challenge for regional associations is their physical distance from the areas where tourism businesses operate (Pechlaner et al., 2012). This distance contributes to a sense of disconnection between businesses and associations, making communication and information flows more challenging. A more effective approach could involve delegating the responsibility for tourism management marketing to local organisations, while still maintaining a holistic understanding of tourist flows and tourism development directions (Garrod & Fyall, 2017). The priority should be on developing tourism, with marketing efforts serving as a supplementary function rather than the primary focus. Moreover, in the current state of associational policy-making, different theoretical approaches are interchangeably replaced without differentiation (Perry, 2007). Since collaboration is often context- and business-specific, reasons to collaborate are difficult criteria to use as targets for policy measures (European Commission, 2003). It is more effective to tackle obstacles for collaboration to mitigate failures (Håkansson & Snehota, 1995; Hoholm, 2015).

For small businesses to become significant stakeholders and overcome barriers to growth, it is crucial to appreciate the collective power that transcends individual capacities. As they face similar challenges, co-located businesses may self-organise to form associations that extend beyond marketing needs. Their shared context fosters common ground and support, enabling local associations to provide a platform for collaboration. In fact, local non-tourism businesses often participate in local associations, motivated not solely by tourism but by a desire to engage in community activities and contribute positively to their local area, despite these associations' goals primarily being focused on tourism. In essence, local associations can be seen as a self-organising outcome of the co-location of businesses that share a sense of community and try to develop the local economy in a meaningful way (Michael, 2007a; Pforr, 2004; Quaranta et al., 2016).

Tourism government bodies must acknowledge the significance of local co-locations. As agglomerations grow, it is essential to recognise that local concentrations of businesses may not always align with administrative borders, leading to the existence of multiple co-located micro-clusters within an agglomeration. Such proximity increases the likelihood of collaboration among tourism businesses due to their shared context and enhanced sense of community, allowing for a more profound understanding of local issues and challenges, which may vary between locations but may also coincide. Empowering local associations and businesses can be pivotal in addressing these challenges and fostering local collaborations. Consequently, tourism development strategies should focus on supporting these clusters rather than merely creating them. Instead of top-down strategies being imposed in a hierarchical structure, local clusters can become primary decision-makers and sources of information for RTOs (Dias-Sardinha et al., 2018).

The role of RTOs should involve providing targeted assistance, resources, and diverse support to address the unique needs and characteristics of different local areas. This approach empowers local clusters to tackle specific issues while also fostering a mechanism for creating an ecosystem and facilitating communication between local clusters, tourism organisations, and the broader environment. By doing so, the focus shifts to local clusters, allowing for better leveraging of the advantages of local co-locations, while also benefiting from the support and resources offered by larger tourism organisations. Essentially, this strategy aims to achieve a more sustainable, community-driven approach to tourism development by empowering and supporting local clusters.

### Partial industrialisation

One of the main challenges in collaboration and related policy initiatives is defining what constitutes tourism. Relying on the notion of tourism businesses as any firm that supplies goods and services to tourists oversimplifies reality, potentially leading to unsuccessful decisions, policies, and collaboration initiatives. Many businesses serving tourists as customers are, in fact, not in the tourism business. Despite having tourists as their customers, some businesses do not identify themselves as tourism businesses since their primary activity lies in producing goods or providing unrelated services. Consequently, these businesses, which often do not consider tourists their main customer base, argue against being classified as tourism enterprises, carrying significant implications for policy and planning. This narrow definition results in a failure to recognise a broader range of businesses that are a part of the tourism ecosystem, emphasising the problem with the conventional approach to tourism as an industry.

This conventional approach is often grounded in the demand side of activities and tourist expenditures, which supposedly represent the tourism industry (Leiper et al., 2008). However, such an understanding reflects a traditional perspective on tourism as a holistic industry. Allocating a firm to the tourism industry based solely on its customer base requires more justification. For example, Leiper (1990) posited that tourism should be considered partially industrialised. In this way, tourism is fully industrialised when businesses are exclusively oriented towards providing goods and services to tourists, implying their offerings are consumed solely by tourists. By contrast, products and services consumed exclusively by non-tourists suggest an opposing condition. Thus, the level of industrialisation indicates a proportion of customers that are tourists, theoretically ranging from fully industrialised tourism to the absence of a tourism industry altogether. Full industrialisation of tourism would imply that the number of destinations is limited to those designated for tourists, requiring formal governance of such destinations. For instance, Western tourist numbers in Bhutan are controlled through visa quotas, bureaucracy procedures, daily tariffs, required guides, and spatial mobility restrictions (Nyaupane & Timothy, 2010). However, instances of fully industrialised tourism are becoming increasingly rare. Under such circumstances, tourists would have fewer travel options globally or within a country.

The concept of partial industrialisation and the resulting divergence of businesses present fundamental challenges for understanding tourism business behaviour with regard to policy and planning. First, the perception of tourism businesses often limits it to businesses directly providing services for tourists. If certain businesses do not see themselves as part of tourism, they are less likely to collaborate with other tourism businesses or participate in RTOs. Understanding these perceptions is crucial for successful collaboration efforts and for developing effective tourism policies and initiatives. Second, there remains a belief that tour operators or agents are the only sources of tourists. This traditional view of tourism, originating from the mass market era, has led regional and national organisations to seek volume by working with tour operators to bring in large groups of tourists. However, the shift from mass tourism to more independently structured travel patterns suggests that tourists increasingly plan their trips independently, with any tourist having the potential to become a customer for a local business. Third, the perception that tourists are solely those who travel from outside a region complicates matters. Leiper's definition of tourists is notably restricted to those traveling outside a region (Leiper, 1992). However, a local family may visit the same attractions, and often pay the same entrance fee as tourists from the other side of the world (with a few exceptions in places where foreigners pay a higher price), and may well also eat in the same restaurant or buy the same souvenirs. On the other side, tourists may visit a local grocery, use a laundry, or go to a local theatre, behaving as a local resident. The perception of tourists is important, especially given the perceived differences between hospitality and tourism businesses (Chapter 3). Notably, UNWTO (Department of Economic Social Affairs, 2017) uses a broad notion of a visitor (domestic, inbound or outbound), which includes a tourist (or overnight visitor) or same-day visitor (excursionist) for statistical purposes. While it may be useful from a statistical point of view, translating this approach to tourism businesses or even RTOs for operational purposes confuses an already complex system. While the perceived market may influence business strategy and marketing, it is unlikely to significantly affect the production processes of goods and services. Consequently, limited understanding of tourism and the narrow perception of tourists can hinder collaboration among businesses. For instance, destination tourism organisations may overlook the tourism value of businesses not traditionally viewed as part of tourism, such as restaurants, factories, and vineyards, thus limiting the potential development of the tourism ecosystem. The importance of perception in shaping tourism collaboration highlights the need to reconsider the traditional approach used to distinguish between different businesses within tourism. The definition of tourism businesses and organisations should be more inclusive, recognising all visitors, including locals, rather than solely focusing on 'tourists'.

Nevertheless, the partial industrialisation perspective is crucial for understanding collaborative behaviour. Only those businesses that believe they are in the tourism business or see tourists as a significant market will likely pursue collaboration with similar or complementary tourism businesses. For example, many businesses, such as grocery stores, retailers, and petrol stations, benefit from tourism and its development. However, they often do not participate in tourism-related associations or collaborate with tourism businesses. These businesses are frequently categorized as 'allied', 'support', 'supplementary', or 'tourism infrastructure' necessary for enhancing the tourist experience, yet they are rarely considered for collaboration. Additionally, some businesses mainly engaged in production activities may not identify as being in the business of tourism, even if tourists make up a portion of their customer base. Moreover, even when organisations like museums acknowledge tourists as part of their market, their community-oriented goals may lead them to view tourism development or marketing organisations as irrelevant. Also, if there are no clear benefits in terms of sales or marketing, businesses that do not significantly rely on the custom of tourists are unlikely to collaborate with tourism businesses or organisations. The common approach of using financial incentives to create a network structure must be approached with caution, as many initiatives fail (Perry, 2007). If there is no common understanding to unify businesses, sustaining collaboration will be difficult, ultimately leading to its failure. Therefore, the success of collaboration partially depends on businesses recognising the role tourism plays in their operations.

### Understanding non-collaborative behaviour

Non-collaborative behaviour must be acknowledged within the existing concepts of clusters, networks, and ecosystems, and policy implications. However, they do not extensively address or explore non-collaborative behaviour in tourism. In the case of clusters, it is essential to acknowledge that geographical proximity can also create tensions and intensify competition (Crick et al., 2020), which may lead to a low level of collaboration (Werner et al., 2015). Network literature often emphasises the positive aspects of networks and their potential for gaining advantages. However, relationships are idiosyncratic and have significant impact on the dynamics of non-collaborative behaviour. The essence of natural ecosystems lies in the complex interactions among living organisms and their environment (Gunderson & Holling, 2001), but non-collaborative behaviour receives limited attention within the tourism ecosystem concept.

Given the similarities between the concepts of clusters, networks, and ecosystems, it is reasonable to argue that non-collaborative behaviour within these concepts is implicitly associated with opportunism. However, opportunistic behaviour often carries a negative connotation (Anderson & Jap, 2005; Ritala & Tidström, 2014; Williamson, 1985), which needs to be addressed in order to realise the collective benefits of collaboration. The costs associated with monitoring and enforcing collaboration are usually higher than coordination norms, as individuals tend to act in their self-interest. While opportunistic behaviour refers mainly to the exploiting of collective efforts for one's own advantage without significant contribution, non-collaborative behaviour is not necessarily negative, as there could be valid reasons for businesses to operate independently.

Although the literature highlights that a lack of trust and trustworthy relationships hinder collaboration (Asero et al., 2017; Raisi et al., 2020), the absence of trust could be attributed to the absence of relationships, perceptions of cultural differences, and/or lack of reciprocity in existing relationships (Hwang & Stewart, 2016; Kim & Shim, 2018). The significance placed on trust in collaborative behaviour often neglects the influence of repeated interactions, reputation, and the capacity to discourage undesired behaviour in the short term (Enright, 1996). Moreover, collaborations that involve low costs may not necessarily demand high-trust relationships. Therefore, while trust can enhance collaboration, it is not the primary determining factor of collaborative behaviour.

Cultural diversity may play a substantial role in determining collaborative and non-collaborative behaviour among businesses. Diversity can come in various forms, including cultural, language, and social differences, and thus creates misunderstandings that hinder collaboration. However, in some cases, such cultural differences may also lead to creative solutions and innovative ideas if a suitable collaboration environment exists for businesses. Therefore, it is important to recognise and understand the impact of cultural differences between actors on collaborative and non-collaborative behaviour, and develop strategies to form a collaborative environment accordingly.

Moreover, small tourism businesses are socially embedded within a locational context (Grabher, 1993), meaning that socio-cultural factors extend to their business behaviour. The exchange of goods and services in an economic context is a specific type of social exchange, and thus similar mechanisms that regulate social behaviour apply to economic behaviour

(Homans, 1961/1974). Such dynamics are particularly evident in the case of closely co-located similar businesses where social and cultural differences play a significant role in collaborative and non-collaborative behaviour. Because of the increased interactions that come with co-location and the social nature of networking, owners' or managers' socio-cultural characteristics significantly impact their behaviour and collaborative decision-making within a local business context. Nevertheless, the shared geographical context holds people together and allows them to co-exist and do business without consensus.

Non-collaborative behaviour is a part of normal business operations. It is significant to recognise that where collaboration exists within tourism business behaviour, non-collaboration also exists. These two behaviours are not separate domains but interconnected aspects of business operations. This interplay becomes especially apparent when considering the spatial context. For example, co-opetition is inherent to co-located businesses in limited geographical proximity. It is not uncommon for businesses in such areas to exhibit non-collaborative behaviour, so it should not be considered abnormal. Constant collaboration is not ideal for fostering innovation, as disruptions are often helpful in encouraging the development of innovative practices. Thus, non-collaborative behaviour can be essential in a competitive environment to drive improvements in products and services, especially when the quality of demand and overall demand is low. Balancing collaboration and non-collaboration in strategies and policies is crucial, acknowledging their coexistence.

## Conclusion

The complexities of tourism collaboration and governance require a re-evaluation of conventional approaches, especially with regard to what measures are used as indicators. The inherent challenges posed by the multitude of actors involved in tourism require effective communication and the establishment of shared objectives. Fostering a collaborative environment where diverse stakeholders can thrive is essential for sustainable growth. Recognising that tourism development and the consequent challenges extend beyond mere economic growth is crucial for creating resilient communities that can adapt to changing circumstances. Understanding regional dynamics and stakeholder roles is paramount for navigating the intricacies of tourism governance. Engaging local businesses enables RTOs to effectively address the specific needs and contexts of different areas. Rethinking the role of RTOs is pivotal in fostering collaboration among tourism businesses. These organisations should focus not only on marketing but also on cultivating shared values and facilitating connections that drive collective growth. Moreover, the challenges of defining the tourism industry highlight the importance of understanding how perceptions influence collaboration among businesses. Acknowledging the diverse contributions of all stakeholders in the tourism ecosystem can lead to more inclusive frameworks that recognise the interconnectedness of various sectors. Fostering a collaborative mindset and integrating multiple perspectives are essential to address the challenges of a rapidly changing world. This evolution ensures that tourism remains a viable and beneficial component of the local and global economies.

# References

Aminudin, N., Kamal, A. S., Jamal, S. A., & Anuar, F. I. (2018). Exploring luxury travel from the perspective of ancillary services supplier: High-end vehicles and limousine service. *International Journal of Supply Chain Management, 7*(5), 443–454.

Anderson, E., & Jap, S. D. (2005). The dark side of close relationships. *MIT Sloan Management Review, 46*(3), 75.

Arzaghi, M., & Henderson, J. V. (2008). Networking off Madison avenue. *The Review of Economic Studies, 75*(4), 1011–1038. 10.1111/j.1467-937x.2008.00499.x

Asero, V., Patti, S., & Skonieczny, S. (2017). Cooperative optimization of tourism networks: An application of a game theory model. In P. Vasant, & M. Kalaivanthan (Eds.), *Handbook of research on holistic optimization techniques in the hospitality, tourism, and travel industry* (pp. 348–364). IGI Global.

Axelrod, R. (2009). *The evolution of cooperation.* Basic Books. (Original work published 1984)

Baggio, R. (2008). Symptoms of complexity in a tourism system. *Tourism Analysis, 13*(1), 1–20.

Baggio, R., Scott, N., & Cooper, C. (2010). Improving tourism destination governance: A complexity science approach. *Tourism Review, 65*(4), 51–60. 10.1108/16605371011093863

Barile, S., Pellicano, M., & Polese, F. (2018). *Social dynamics in a systems perspective.* Springer Cham. 10.1007/978-3-319-61967-5

Barile, S., & Saviano, M. (2018). Complexity and sustainabi lity in management: Insights from a systems perspective. In S. Barile, M. Pellicano, & F. Polese (Eds.), *Social dynamics in a systems perspective* (pp. 39–63). Springer Cham. 10.1007/978-3-319-61967-5_3

Beaumont, N., & Dredge, D. (2010). Local tourism governance: A comparison of three network approaches. *Journal of Sustainable Tourism, 18*(1), 7–28. 10.1080/09669580903215139

Bengtsson, M., & Kock, S. (2014). Coopetition – Quo vadis? Past accomplishments and future challenges. *Industrial Marketing Management, 43*(2), 180–188. 10.1016/j.indmarman.2014.02.015

Bush, G. (2006). Learning about learning: From theories to trends. *Teacher Librarian, 34*(2), 14–19.

Butcher, J. (2021). Covid-19, tourism and the advocacy of degrowth. *Tourism Recreation Research, 48*(5), 633–642. 10.1080/02508281.2021.1953306

Chapman, G. (2009). Chaos and complexity. In R. Kitchin & N. Thrift (Eds.), *International encyclopedia of human geography* (pp. 31–39). Elsevier. 10.1016/B978-008044910-4.00412-0

Crick, J. M., Crick, D., & Tebbett, N. (2020). Competitor orientation and value co-creation in sustaining rural New Zealand wine producers. *Journal of Rural Studies, 73,* 122–134. 10.1016/j.jrurstud.2019.10.019

Curry, O., & Dunbar, R. I. M. (2013). Do birds of a feather flock together? *Human Nature, 24*(3), 336–347. 10.1007/s12110-013-9174-z

Department of Economic Social Affairs. (2017). *International recommendations for tourism statistics 2008.* United Nations. 10.18356/791169b3-en

Dias-Sardinha, I., Ross, D., & Calapez Gomes, A. (2018). The clustering conditions for managing creative tourism destinations: The Alqueva region case, Portugal. *Journal of Environmental Planning and Management, 61*(4), 635–655. 10.1080/09640568.2017.1327846

Enright, M. J. (1996). Regional clusters and economic development: A research agenda. In U. H. Staber, N. V. Schaefer, & B. Sharma (Eds.), *Business networks: Prospects for regional development* (pp. 190–214). De Gruyter. 10.1515/9783110809053.190

Enright, M. J., & Newton, J. (2004). Tourism destination competitiveness: A quantitative approach. *Tourism Management, 25*(6), 777–788. 10.1016/j.tourman.2004.06.008

European Commission. (2003). *The observatory for SMEs: SMEs and cooperation* (Fifth Report). Publications Office.

Farrell, B. H., & Twining-Ward, L. (2004). Reconceptualizing tourism. *Annals of Tourism Research, 31*(2), 274–295. 10.1016/j.annals.2003.12.002

Faulkner, B., & Russell, R. (2003). Chaos and complexity in tourism: In search of a new perspective. In F. Liz, K. J. Leo, & C. Chris (Eds.), *Progressing tourism research - Bill Faulkner* (pp. 205–219). Channel View Publications. 10.21832/9781873150498-015

Felzensztein, C., Gimmon, E., & Deans, K. R. (2018). Coopetition in regional clusters: Keep calm and expect unexpected changes. *Industrial Marketing Management, 69,* 116–124. 10.1016/j.indmarman.2018.01.013

Furnham, A. (1988). *Lay theories: Everyday understanding of problems in the social sciences* (Vol. 15). Pergamon. 10.1016/C2009-0-14697-4

Fyall, A., Garrod, B., & Wang, Y. (2012). Destination collaboration: A critical review of theoretical approaches to a multi-dimensional phenomenon. *Journal of Destination Marketing & Management, 1*(1–2), 10–26. 10.1016/j.jdmm.2012.10.002

Garrod, B., & Fyall, A. (2017). Collaborative destination marketing at the local level: Benefits bundling and the changing role of the local tourism association. *Current Issues in Tourism, 20*(7), 668–690. 10.1080/13683500.2016.1165657

Gleick, J. (1987). *Chaos: Making a new science.* Penguin.

Gössling, S., Ring, A., Dwyer, L., Andersson, A. C., & Hall, C. M. (2016). Optimizing or maximizing growth? A challenge for sustainable tourism. *Journal of Sustainable Tourism, 24*(4), 527–548. 10.1080/09669582.2015.1085869

Gössling, S., Scott, D., & Hall, C. M. (2021). Pandemics, tourism and global change: A rapid assessment of COVID-19. *Journal of Sustainable Tourism, 29*(1), 1–20. 10.1080/09669582.2020.1758708

Grabher, G. (1993). *The Embedded firm: On the socioeconomics of industrial networks.* Routledge.

Gray, B. (1989). *Collaborating: Finding common ground for multiparty problems.* Jossey-Bass.

Gunderson, L. H., & Holling, C. S. (2001). *Panarchy: Understanding transformations in human and natural systems.* Island Press.

Håkansson, H., & Snehota, I. (1995, September). *The burden of relationships or who's next. IMP Conference (11th), Manchester.*

Hall, C. M. (1999). Rethinking collaboration and partnership: A public policy perspective. *Journal of Sustainable Tourism, 7*(3–4), 274–289. 10.1080/09669589908667340

Hill, R. A., & Dunbar, R. I. M. (2003). Social network size in humans. *Human Nature, 14*(1), 53–72. 10.1007/s12110-003-1016-y

Hoholm, T. (2015). Interaction avoidance in networks. *IMP Journal, 9*(2), 117–135. 10.1108/IMP-03-2015-0011

Homans, G. C. (1974). *Social behavior: Its elementary forms* (Revised edition). Harcourt, Brace, Jovanovich. (Original work published 1961)

Hwang, D., & Stewart, W. P. (2016). Social capital and collective action in rural tourism. *Journal of Travel Research, 56*(1), 81–93. 10.1177/0047287515625128

Kallmuenzer, A., Zach, F. J., Wachter, T., Kraus, S., & Salner, P. (2021). Antecedents of coopetition in small and medium-sized hospitality firms. *International Journal of Hospitality Management, 99.* 10.1016/j.ijhm.2021.103076

Kelliher, F., Reinl, L., Johnson, T. G., & Joppe, M. (2018). The role of trust in building rural tourism micro firm network engagement: A multi-case study. *Tourism Management, 68*, 1–12. 10.1016/j.tourman.2018.02.014

Kesić, S. (2016). Systems biology, emergence and antireductionism. *Saudi Journal of Biological Sciences, 23*(5), 584–591. 10.1016/j.sjbs.2015.06.015

Kim, N., & Shim, C. (2018). Social capital, knowledge sharing and innovation of small- and medium-sized enterprises in a tourism cluster. *International Journal of Contemporary Hospitality Management, 30*(6), 2417–2437. 10.1108/ijchm-07-2016-0392

Kylänen, M., & Rusko, R. (2011). Unintentional coopetition in the service industries: The case of Pyhä-Luosto tourism destination in the Finnish Lapland. *European Management Journal, 29*(3), 193–205. 10.1016/j.emj.2010.10.006

Leiper, N. (1979). The framework of tourism: Towards a definition of tourism, tourist, and the tourist industry. *Annals of Tourism Research, 6*(4), 390–407. 10.1016/0160-7383(79)90003-3

Leiper, N. (1990). Partial industrialization of tourism systems. *Annals of Tourism Research, 17*(4), 600–605. 10.1016/0160-7383(90)90030-U

Leiper, N. (1992). *Whole tourism systems: Interdisciplinary perspectives on structures, functions, environmental issues and management.* [Doctoral thesis, Massey University].

Leiper, N., Stear, L., Hing, N., & Firth, T. (2008). Partial industrialisation in tourism: A new model. *Current Issues in Tourism, 11*(3), 207–235. 10.2167/cit356.0

Letaifa, S. B., & Rabeau, Y. (2013). Too close to collaborate? How geographic proximity could impede entrepreneurship and innovation. *Journal of Business Research, 66*(10), 2071–2078. 10.1016/j.jbusres. 2013.02.033

López-Sanz, J. M., Penelas-Leguía, A., Gutiérrez-Rodríguez, P., & Cuesta-Valiño, P. (2021). Sustainable development and consumer behavior in rural tourism – The importance of image and loyalty for host communities. *Sustainability, 13*(9). 10.3390/su13094763

Lorenz, E. N. (1963). Deterministic nonperiodic flow. *Journal of Atmospheric Sciences, 20*(2), 130–141. 10.1175/1520-0469(1963)020 <0130:DNF>2.0.CO;2

Lorenz, E. N. (1993). *The essence of chaos*. University of Washington Press.

Markusen, A. (2003). Fuzzy concepts, scanty evidence, policy distance: The case for rigour and policy relevance in critical regional studies. *Regional Studies, 37*(6–7), 701–717. 10.1080/0034340032000108796

Michael, E. J. (2007a). *Micro-clusters and networks: The growth of tourism*. Routledge. 10.4324/9780080464909

Michael, E. J. (2007b). Micro-clusters in tourism. In E. J. Michael (Ed.), *Micro-clusters and networks: The growth of tourism* (pp. 33–42). Routledge.

Mill, R. C., & Morrison, A. M. (1985). *The tourism system: An introductory text*. Prentice-Hall International.

Mills, J. A. (1998). *Control: A history of behavioral psychology*. New Yor University Press.

Moore, J. F. (1993). Predators and prey: A new ecology of competition. *Harvard Business Review, 71*(3), 75–86.

Nishimura, J., & Okamuro, H. (2011). Subsidy and networking: The effects of direct and indirect support programs of the cluster policy. *Research Policy, 40*(5), 714–727. 10.1016/j.respol.2011.01.011

Nyaupane, G. P., & Timothy, D. J. (2010). Power, regionalism and tourism policy in Bhutan. *Annals of Tourism Research, 37*(4), 969–988. 10.1016/j. annals.2010.03.006

Pechlaner, H., Raich, F., & Kofink, L. (2011). Elements of corporate governance in tourism organizations. *Tourismos, 6*(3), 57–76. 10.26215/tourismos.v6i3.248

Pechlaner, H., Volgger, M., & Herntrei, M. (2012). Destination management organizations as interface between destination governance and corporate governance. *Anatolia, 23*(2), 151–168. 10.1080/13032917. 2011.652137

Perry, M. (2007). From networks to clusters and back again: A decade of unsatisfied policy aspiration in New Zealand. In R. MacGregor & A. Hodgkinson (Eds.), *Small business clustering technologies: Applications in marketing, management, IT and economics* (pp. 160–183). IGI Global.

Pforr, C. (2004). Policy-making for sustainable tourism. In F. D. Pineda & C. A. Brebbia (Eds.), *Sustainable Tourism* (pp. 83–94). WIT Press.

Phillipson, J., Gorton, M., & Laschewski, L. (2006). Local business co-operation and the dilemmas of collective action: Rural micro-business networks in the North of England. *Sociologia Ruralis, 46*(1), 40–60. 10.1111/j.1467-9523.2006.00401.x

Pike, S., & Page, S. J. (2014). Destination marketing organizations and destination marketing: A narrative analysis of the literature. *Tourism Management, 41*, 202–227. 10.1016/j.tourman.2013.09.009

Pilving, T., Kull, T., Suškevics, M., & Viira, A. H. (2022). Diverse networks in regional tourism: Ties that foster and hinder the development of rural-urban tourism collaboration in estonia. *European Journal of Tourism Research, 30*. 10.54055/ejtr.v30i.2596

Polese, F. (2018). Successful value co-creation exchanges: A VSA contribution. In S. Barile, M. Pellicano, & F. Polese (Eds.), *Social dynamics in a systems perspective* (pp. 39–63). Springer Cham. 10.1007/978-3-319-61967-5_2

Porter, M. E. (1990). *The competitive advantage of nations.* Macmillan.

Pyke, F., Becattini, G., & Sengenberger, W. (1990). *Industrial districts and inter-firm co-operation in Italy.* International Institute for Labour Studies.

Quaranta, G., Citro, E., & Salvia, R. (2016). Economic and social sustainable synergies to promote innovations in rural tourism and local development. *Sustainability, 8*(7), 668. 10.3390/su8070668

Raisi, H., Baggio, R., Barratt-Pugh, L., & Willson, G. (2020). A network perspective of knowledge transfer in tourism. *Annals of Tourism Research, 80*, 102817. 10.1016/j.annals.2019.102817

Ritala, P., & Tidström, A. (2014). Untangling the value-creation and value-appropriation elements of coopetition strategy: A longitudinal analysis on the firm and relational levels. *Scandinavian Journal of Management, 30*(4), 498–515. 10.1016/j.scaman.2014.05.002

Roberts, S. G. B., Dunbar, R. I. M., Pollet, T. V., & Kuppens, T. (2009). Exploring variation in active network size: Constraints and ego characteristics. *Social Networks, 31*(2), 138–146. 10.1016/j.socnet.2008.12.002

Schrank, A., & Whitford, J. (2011). The anatomy of network failure. *Sociological Theory, 29*(3), 151–177. 10.1111/j.1467-9558.2011.01392.x

Scott, N., Baggio, R., & Cooper, C. (2008). *Network analysis and tourism: From theory to practice.* Channel View Publications.

Sigurðardóttir, I., & Steinthorsson, R. S. (2018). Development of micro-clusters in tourism: A case of equestrian tourism in northwest Iceland. *Scandinavian Journal of Hospitality and Tourism, 18*(3), 261–277. 10.1080/15022250.2018.1497286

Speakman, M. (2017). A paradigm for the twenty-first century or meta-phorical nonsense? The enigma of complexity theory and tourism research. *Tourism Planning & Development, 14*(2), 282–296. 10.1080/21568316.2016.1155076

Stiglitz, J. E. (2009). GDP fetishism. *The Economists' Voice, 6*(8). 10.2202/1553-3832.1651

Tronvoll, B., Barile, S., & Caputo, F. (2018). A systems approach to understanding the philosophical foundation of marketing studies. In S.

Barile, M. Pellicano, & F. Polese (Eds.), *Social dynamics in a systems perspective* (pp. 1–18). Springer Cham. 10.1007/978-3-319-61967-5_1

Tunisini, A., & Marchiori, M. (2020). Why do network organizations fail? *Journal of Business & Industrial Marketing, 35*(6), 1011–1021. 10.1108/JBIM-01-2019-0056

UNWTO. (2023). *International tourism highlights, 2023 edition – The impact of COVID-19 on tourism (2020–2022)*. 10.18111/9789284424986

Wang, Y., & Krakover, S. (2008). Destination marketing: Competition, cooperation or coopetition? *International Journal of Contemporary Hospitality Management, 20*(2), 126–141. 10.1108/09596110810852122

Werner, K., Dickson, G., & Hyde, K. (2015). Coopetition and knowledge transfer dynamics: New Zealand's regional tourism organizations and the 2011 Rugby World Cup. *Event Management, 19*. 10.3727/15259951 5X14386220874841

Wesley, A., & Pforr, C. (2010). The governance of coastal tourism: Unravelling the layers of complexity at Smiths beach, Western Australia. *Journal of Sustainable Tourism, 18*(6), 773–792. 10.1080/0966958 1003721273

Wijesinghe, S. N. R. (2022). Neoliberalism, Covid-19 and hope for transformation in tourism: The case of Malaysia. *Current Issues in Tourism, 25*(7), 1106–1120. 10.1080/13683500.2021.2012431

Williamson, O. E. (1985). *The economic institutions of capitalism: Firms, markets, relational contracting.* Free Press.

Zahra, A., & Ryan, C. (2007). From chaos to cohesion – complexity in tourism structures: An analysis of New Zealand's regional tourism organizations. *Tourism Management, 28*(3), 854–862. 10.1016/j.tourman.2006.06.004

Zhong, Q., & Tang, T. (2018). Impact of government intervention on industrial cluster innovation network in developing countries. *Emerging Markets Finance and Trade, 54*(14), 3351–3365. 10.1080/1540496X.2018.1434504

Zhou, W.-X., Sornette, D., Hill, R. A., & Dunbar, R. I. M. (2005). Discrete hierarchical organization of social group sizes. *Proceedings of the Royal Society B: Biological Sciences, 272*(1561), 439–444. 10.1098/rspb.2004.2970

Ziakas, V. (2024). *Rethinking events: A critique and reconfiguration.* Edward Elgar Publishing. 10.4337/9781035313648

# 5    Conclusions

## Introduction

Tourism represents multifaceted and complex economic activity. Strategic collaboration and effective governance emerge as crucial elements for facilitating resilience, adaptability, and sustainable development. Tourism is shaped by a diverse array of businesses, each facing unique constraints and opportunities, particularly in balancing day-to-day operational demands with collaborative behaviours. For small-sized tourism businesses, varying in size, structure, and management approach, engagement in collaborations can be challenging due to limited resources. Although these constraints do not necessarily indicate a lack of willingness to collaborate, they do indicate that small businesses may face more barriers to collaboration than their larger counterparts. Smaller enterprises often employ a utilitarian approach, weighing the costs and benefits of engagement. Feedback from past collaborative experiences greatly impacts whether businesses sustain, enhance, or withdraw from collaborations and associations, with perceived low value significantly contributing to non-collaborative behaviours.

The governance of tourism requires adaptable frameworks that respect not only economic growth but also the complex relational and geographical dimensions of everything, from local clusters to networks and expansive ecosystems. Clusters allow local businesses to strengthen interactions and capitalise on proximity, while networks enable connections beyond immediate locales, forming expansive ecosystems of collaboration. Engaging stakeholders across this spectrum allows tourism businesses and organisations to build and sustain meaningful collaborations. Acknowledging the differences in form and function among clusters, networks, and ecosystems is essential to understanding the landscape of tourism collaboration and governance.

Being a complex system, the associational tourism economy evolves and adapts over time, while collaboration serves as a crucial mechanism for sustained collective development. However, effective collaboration

DOI: 10.4324/9781003293606-5

extends beyond basic coordination and cooperation. It requires purposeful alignment of visions, strategic evaluation of internal capacities, and regular, structured networking that supports individual and mutual growth and adaptability. This approach encourages both competition and cooperation, creating a tourism ecosystem where diverse stakeholders, whether large or small, local or regional, can continue to be a resilient and impactful component of local and global tourism economy.

## The role of tourism associations

Government-led and business-led tourism associations play an essential role in tourism development by facilitating knowledge exchange, best practice sharing, joint marketing campaigns, and collaborative projects that strengthen tourism systems. To support sustainable tourism development, these associations must extend their focus beyond attracting new members and marketing regions to acknowledge the multi-layered nature of tourism itself. Understanding the factors that influence collaborative behaviour can help regional tourism associations to develop strategies and interventions that create a collaborative environment and mindset among stakeholders.

While marketing is indeed critical for promoting destinations and attracting visitors, an overemphasis on it can sometimes belittle essential aspects of effective and sustainable tourism management. Regional tourism organisations' responsibilities often extend far beyond marketing to include, for example, destination and infrastructure planning, stakeholder engagement, or the implementation of sustainable practices. When marketing is prioritised, there is a risk of neglecting the management and planning functions essential to tourism's sustainability. A more balanced approach that integrates both marketing and management priorities would enable regional tourism organisations to better address the varied needs and challenges of local destinations, fostering sustainable tourism development and generating lasting positive social and economic impacts for both tourists and local communities.

Tourism associations have increasingly leaned towards the commercialisation of their functions, prioritising profit-oriented objectives. Although associations need to show their value and generate returns on taxpayer money, this shift in focus may undermine their roles as facilitators of collaboration and sustainable tourism ecosystems. These organisations must carefully balance the imperatives to justify their funding within the broader goals of tourism development and community well-being. Given their resource constraints and the pressure to demonstrate financial viability, tourism associations should thoughtfully consider which performance indicators they adopt. Emphasising long-term tourism development objectives is crucial, as commercialising tourism associations' functions can undermine efforts towards sustainable tourism.

Delegating tourism development responsibilities to local associations while retaining a comprehensive perspective on tourist flows and development direction can enhance the effectiveness of these associations. This decentralised approach empowers local businesses and communities to take ownership of tourism development, fostering local pride and collaboration within local clusters. To remain effective intermediaries, associations should align their activities to bridge the gaps between businesses, government agencies, and local communities. While associations play a pivotal role in tourism, they must carefully manage their commercial orientation. Adopting an inclusive and collaborative approach allows associations to address the diverse needs and challenges faced by tourism businesses, and to support the long-term success and resilience of tourism development.

## Geographical proximity and dynamics of collaboration

Geographical proximity can be understood in two ways, as operational and physical co-location. Operational proximity refers to businesses sharing a common operational space or area, while physical proximity refers to their actual physical locations. Both types of proximity play crucial roles in shaping collaborative and non-collaborative behaviours. The spatial context creates opportunities for frequent face-to-face interactions, which enhance mutual awareness. This increased awareness contributes to a sense of community, promoting shared norms and values, which in turn may facilitate collaboration, raising the potential for business's competitiveness. To fully leverage the benefits of geographical proximity, stakeholders should create an inclusive and supportive environment that encourages a collaborative mindset. Tourism businesses, in turn, must adopt an open approach towards collaboration to maximise the potential of close proximity.

However, it is important to acknowledge that geographical proximity alone does not guarantee collaboration. It is equally important to recognise the existence of non-collaborative behaviour within spatial contexts. Differences in business priorities, objectives, and competitive strategies may contribute to this behaviour. Collaboration and non-collaboration are not mutually exclusive, as they coexist and intertwine in influencing the overall development process. Rather than focusing solely on promoting collaboration, it is essential to appreciate the role of non-collaboration and its potential positive contributions. In certain contexts, non-collaborative behaviour can be essential in driving improvements, fostering innovation, and responding to competitive pressures. Non-collaboration stimulates businesses to seek independent solutions, explore new strategies, and differentiate themselves in the market, promoting individual growth. A more holistic approach that recognises the coexistence and interdependence of these behaviours to find the right balance between fostering collaborations and addressing non-collaborations can guide policymakers and industry practitioners in fostering a balanced environment. While collaboration is often regarded as a positive and desirable activity, non-collaboration also plays a significant role in the development of tourism systems and should not be overlooked. Both collaborative and non-collaborative behaviours can contribute constructively to the development of tourism, enhancing the sustainability and resilience of local economies.

A well-functioning tourism system must acknowledge not only economic factors but also socio-cultural and environmental considerations, fostering responsible tourism practices, preserving natural and cultural heritage, and engaging local communities. To achieve this, it is essential to explore the underlying motivations and reasons behind both collaboration and non-collaboration, examine their implications for the overall functioning of tourism systems, and develop strategies to mitigate the adverse effects on destination communities.

## Suggestions for enhancing collaboration

Collaboration is one of the strategies businesses can utilise, and small businesses in particular have to understand their needs at different stages of their development to leverage the benefits of collaboration when it aligns with those needs. The smaller size of tourism businesses often means they have limited resources, but this does not imply that they are unsuccessful or unwilling to collaborate. Rather, the operational and strategic requirements of small tourism businesses may not always align with the outcomes typically associated with collaborative strategies. As a result, the perceived benefits from collaboration might not be sufficient to justify the expenses and efforts required for such collaboration. Despite this, it is crucial for small businesses to approach networking structurally and remain open to collaboration opportunities when they align with their specific needs and development stages. Adopting an open, collaborative mindset will allow small businesses to effectively evaluate and leverage the potential benefits of collaboration, helping them maximise its value.

Tourism associations and businesses should focus on creating mutual value and benefits through their collaborative efforts. Understanding the specific needs and expectations of potential counterparts is essential, allowing for tailored collaborative strategies that align with perceived value. This process involves identifying shared goals, establishing clear communication channels, and developing mutually beneficial incentives that foster collaboration. Table 5.1 outlines key areas of collaboration between tourism businesses to consider, providing suggestions to foster individual and mutual benefits.

The influence of geographical proximity on collaboration should not be underestimated. Close proximity between businesses creates opportunities for frequent face-to-face interactions, networking, and knowledge

*Table 5.1* Enhancing business-to-business collaboration

| Areas | Potential actions |
| --- | --- |
| Explore opportunities | Expand collaborations beyond the traditional transportation, accommodation, and activity chains to uncover new opportunities in the area |
| Collaborative training programmes | Pool resources with other businesses to organise training programmes or workshops that benefit all participants and improve the overall quality of tourism services in the region |
| Development focus | Emphasise not only mutual marketing but also the management of local tourism and efforts to mitigate negative consequences |
| Data exchange | Engage in relevant data exchange among local businesses to identify market trends, tourist preferences, and growth opportunities |
| Community engagement | Collaborate with local communities and community-based organisations to create authentic and sustainable tourism experiences that benefit both tourists and residents |
| Resource sharing | Examine opportunities for sharing with other tourism businesses such as, for example, hours of work, facilities, and staff that can be especially helpful during shoulder seasons |
| Flexibility and adaptability | Create contingency plans to ensure flexibility and adaptability during unforeseen challenges, considering the changing market conditions and dynamic nature of tourism |
| Explore digital solutions | Examine a wide range of digital tools to create seamless experiences for tourists, as well as enhancing communication, knowledge sharing, and collaboration among businesses |
| Conflict resolution | Integrate compelling conflict resolution mechanisms/ procedures among businesses to address and resolve any misunderstandings, challenges, and conflicts that may arise |
| Sustainable practices | Emphasise sustainable practices to ensure the local destination's long-term environmental and socio-cultural benefits |
| Research | Collaborate with research institutions and universities on joint research and development projects that contribute to the local area's tourism development |
| Monitoring and evaluation | Implement a monitoring and evaluation framework to track the progress and impact of collaborations, enabling data-driven decisions for continuous improvement |
| Commit long-term | View collaborations as long-term relationships rather than short-term transactions with immediate benefits |

exchange, which can significantly enhance collaborative efforts. Local organisations need not confine businesses within strict membership structures but should facilitate the free flow of information and best practices, supporting continuous service improvement and innovation, especially in rural tourism destinations. Organising local events, workshops, or networking platforms can encourage connections among businesses within specific geographical areas, fostering a sense of community and shared purpose. While marketing is important for tourism organisations, local businesses may seek a broader range of outcomes, such as knowledge sharing, innovation, and the creation of a sustainable collaborative environment. Establishing local organisations that focus on building collective strength and internal cohesion, rather than solely on marketing the destination, can be particularly beneficial. Table 5.2 highlights key areas for businesses to consider in their collaboration with associations, emphasising the importance of exploring diverse associations, actively participating, and contributing to the development of sustainable and innovative tourism practices. These strategies help foster a dynamic, knowledge-sharing environment in local tourism communities.

Although professional and regional tourism associations play a pivotal role in shaping the collaborative environment, they often face significant challenges and missed opportunities. As associations grow in size, their ability to meet the specific needs of individual businesses diminishes, as does their capacity to address local concerns. The pursuit of representing the entire industry can sometimes compromise their focus on the unique challenges of smaller, geographically distinct businesses. To counter this,

*Table 5.2* Enhancing business-to-association collaboration

| Areas | Potential actions |
| --- | --- |
| Explore diverse associations | Explore a broader network of associations for potential collaboration and membership opportunities as associations differ in their offerings |
| Participate actively | Actively engage with tourism associations by attending meetings, events, and networking opportunities. Establishing a strong presence within associations demonstrates a commitment to collaborative efforts |
| Provide feedback | Offer constructive feedback to shape future initiatives and strengthen collaborations by initiating regular open dialogues and communication with existing tourism associations |
| Contribute | Contribute insights and knowledge with associations, adding value to the discussions and initiatives |
| Explore funding opportunities | Explore funding opportunities from various organisations to support tourism development in a local area |
| Establish independent association | Form a separate and self-governing organisation to serve the interests of local businesses and facilitate communication with regional tourism organisations |

regional tourism organisations must adopt a broader perspective, recognising the diverse needs of localities and prioritising initiatives that foster a supportive environment for networking, collaboration, and knowledge exchange within and between these areas. Despite the growing emphasis on marketing well-known destinations and large businesses within them, it is critical that regional tourism organisations focus on comprehensive destination management plans, recognising the geographical context of business operations. Table 5.3 outlines key areas for associations to focus on to enhance their role in tourism development.

From a policy perspective, it is vital to consider the specific needs of micro- and medium-sized tourism businesses when developing policies and support initiatives. For example, regional tourism associations and governments should recognise the diversity in business sizes and the cultural backgrounds of owners, including limited networking opportunities

*Table 5.3* Enhancing collaboration within associations

| Areas | Potential actions |
| --- | --- |
| Consider tourism planning and development | Balance marketing with the management of tourism development, as focusing only on marketing can present challenges in achieving sustainable growth |
| Establish channels of communication | Establish regular communication channels with local businesses to encourage feedback, open dialogue, and flow of ideas |
| Digital technology integration | Embrace integrating digital tools and platforms to enhance communication, foster knowledge sharing, and facilitate collaboration between local associations and organisations |
| Promote and empower local organisations | Support the establishment of local organisations to encourage community engagement and develop tourism at their own pace, addressing their specific needs |
| Cross-sector collaborations | Explore collaboration opportunities among businesses and associations from different tourism sectors to facilitate access to diverse expertise and perspectives |
| Peer support/ mentorship programmes | Develop mentorship and peer support programmes that facilitate connections between experienced businesses and newcomers to promote knowledge transfer and skill development |
| Research collaborations | Conduct regular research integration to foster innovation, service, and product development in tourism, extending beyond marketing research alone |
| Crisis response and recovery | Develop comprehensive, collaborative crisis response and recovery plans for businesses and associations to support each other during adversities, including communication with the government on essential needs |
| Evaluate and adapt | Continuously assess collaborative effectiveness, review feedback, and remain flexible to adapt to changing circumstances |

and the need for tailored policy support. Policies and resources should be adapted to ensure equal accessibility and understanding, allowing businesses of all sizes to engage meaningfully in collaborations. Local governments, in particular, need to play a more active role in fostering strong relationships within the local business community, as the long-term competitiveness of the tourism offering heavily depends on the success and resilience of small- and medium-sized enterprises. The creation of value through collaboration, the recognition of geographical factors, and the tailored support for diverse business needs are key to fostering a sustainable and competitive tourism environment.

## Closing thoughts

This book serves as an introduction to the complexities of collaboration in tourism, primarily focusing on the geographical proximity of businesses and the importance of collaboration through concepts of clusters, networks, and ecosystems. While tourism networks, clusters, and ecosystems have long been part of tourism development, there is still much to learn about their interconnectivity and how they can be more effectively leveraged to enhance collaboration. As we look towards the future of tourism and collaboration, it is clear that the path forward will require innovative thinking, cross-sector collaboration, and a deep understanding of the dynamics shaping the associational tourism economy. Collaboration, both within and beyond tourism, will be fundamental in addressing the complex challenges we face, from environmental sustainability to socio-economic equity. While the book offers a foundational understanding, it is by no means exhaustive. Instead, the aim is to provide readers with a starting point for further exploration into collaborative behaviour in tourism by highlighting the multifaceted nature of tourism systems and the complex collaborative dynamics that drive them.

   While the book provides a foundation, it is the continued engagement of academics, practitioners, policymakers, and communities that will drive meaningful change and foster a more sustainable, resilient, and inclusive tourism for the future.

   Tourism is a captivating field of study, research, and practice, offering insights into both local and global economies, as well as into human behaviour and societal change. I hope this book encourages readers to think critically, engaging in a more reflective and analytical way, and to seek out additional perspectives, challenge assumptions, and build upon the ideas presented here.

   Thank you for engaging with this book. I wish you all the best in your ongoing exploration of tourism, and in your future endeavours in this ever-evolving field.

# Index